Abstract differential equations

S D Zaidman
Université de Montreal

Abstract differential equations

Pitman Advanced Publishing Program
SAN FRANCISCO·LONDON·MELBOURNE

PITMAN PUBLISHING LIMITED
39 Parker Street, London WC2B 5PB

North American Editorial Office
1020 Plain Street, Marshfield, Massachusetts 02050

North American Sales Office
FEARON PITMAN PUBLISHERS INC.
6 Davis Drive, Belmont, California 94002

Associated Companies
Copp Clark Pitman, Toronto
Pitman Publishing New Zealand Ltd, Wellington
Pitman Publishing Pty Ltd, Melbourne

© S D Zaidman 1979

AMS Subject Classifications: (main) 34G05
(subsidiary) 47E05, 47-XX

Library of Congress Cataloging in Publication Data

Zaidman, Samuel, 1933-
    Abstract differential equations.

    (Research notes in mathematics ; 36)
    1. Differential equations. 2. Cauchy problem.
I. Title. II. Series.
QA371.Z2   515'.35   79-9061
ISBN 0-8224-8427-7

Manufactured in Great Britain

US ISBN 0-8224-8427-7
UK ISBN 0 273 08427 5

Preface

The subject matter which is covered in this monograph deals with linear differential equations in Banach and Hilbert spaces whose "coefficients" are linear unbounded operators.

When we try to study and understand this topic we feel that we are working in ordinary differential equations; however, the discontinuity of the operator coefficients makes things to look quite different from what is known in the classical case.

When we seek for examples or applications we find the most natural ones in the theory of partial differential equations: if to one of the variables is given a privilegiate position and all the other are put together we obtain "at once" an ordinary "differential" equation with respect to the variable which was put in evidence; thus for example, the heat or the wave equation, and even the -non of evolution type- Laplace's equation give rise to ordinary differential equations of this kind. Adding boundary conditions in order to ensure definiteness of the solutions can often be translated in terms of considering solutions in some convenient linear-often normed- function spaces.

Therefore, we can frequently guess or even prove theorems on differential equations in Banach spaces looking at a corresponding pattern in partial differential equations; later on, we may develop a theory of "abstract" differential equations in an independent way and see from time to time what can be obtained when returning to the more classical problems.

I shall give now a few, rather personal, "historical" information:

Professor E. HILLE, in several papers and in his famous book on "Functional Analysis and Semi-Groups" [15] is the founder of this theory; I have been introduced to this brilliantly written and highly original work - in the Russian translation - by Professor A. Halanay to whom my warmest thanks are due.

(Main contributors to the study of abstract Cauchy problems were K. Yosida (the now classical "Hille-Yosida" theorem), R. Phillips, I. Miyadera, S. G. Krein, to mention only a few names).

Later on Professors S. Agmon and L. Nirenberg helped me to enter in the wide world of their (now classical) article "Properties of solutions of ordinary differential equations in Banach spaces" (Comm. Pure Appl. Math. XVl (1963), pp. 121-239) and its influence on the present monograph could hardly be overestimated.

Seminar talks on some topics covered by our actual work were delivered at the University of Bucharest (in the 1950s) and at the Politecnico di Milano (in 1961-62-63); then, at the "Séminaire de Mathématiques Supérieures, Université de Montreal", Summer 1965. An almost complete account of those lectures were printed at the Politecnico di Milano (a personal edition, by myself, with help of Professor C. Vaghi, M. L. Ricci-Boella, A. Vasconi-Airoldi), and by the University of Montreal press in 1966.

The appearance of a series of important new monographs on this topic (see [7], [11], [20], [21], [29]) indicates the necessity to keep in contact with the newest developments. I thought that it could be useful if I would present to the mathematical community an enlarged and revised version of my Montreal 1965 lecture notes; the result of my work is presented in the actual monograph.

My warmest thanks are due to Professor L. PAYNE (Cornell University) who accepted my Notes for this series; to the Editors, Dr. B. Weghofer and

S. K. Hemmings, for their warm support and encouragement; to Miss T. Ouellet

who typed my handwritten (often clumsy) manuscript; and finally to the

National Research Council of Canada (now the Natural Sciences and Engineering

Research Council of Canada) whose financial support extended over several

years was quite essential in the preparation of this work.

Contents

1 The Cauchy problem and associated semi-groups

INTRODUCTION

In this chapter we remember the definition of the Cauchy problem for linear systems of n differential equations with n unknowns; then we pass to a more general situation, which could be described, roughly as a study of the same problem for infinite systems of differential equations with an infinity of unknowns. We define, following a fundamental monograph by S. G. Krein [20], the well-posed Cauchy problem for a linear differential equation in a Banach space; the naturally associated semi-group of operators is introduced and its continuity properties are examined, as well as its infinitesimal generator.

§1. THE EUCLIDEAN CASE

The differential equation

$$\frac{dx(t)}{dt} = ax(t) \quad , \quad a \text{ being a constant} \tag{1.1}$$

is "the simplest" differential equation; it admits as solutions all the exponential functions: $K \exp(at)$ where K is an arbitrary real number.

The Cauchy problem for (1.1) is obtained by adding to (1.1) an "initial" condition; if t_o is an arbitrary real number we shall seek among all solutions of (1.1) those verifying in addition the equality:

$$x(t_o) = x_o \quad , \quad \text{where } x_o \text{ is a given real number.} \tag{1.2}$$

This choice has as an obvious consequence that

$$x(t) = \exp(a(t - t_o))x_o \tag{1.3}$$

which is the unique solution of the problem (1.1) - (1.2).   This

situation is generalized by considering a set of n differential equations

$$\frac{dx_1}{dt} = a_{11}x_1 + a_{12}x_2 \cdots a_{1n}x_n$$

$$\frac{dx_2}{dt} = a_{21}x_1 + a_{22}x_2 \cdots a_{2n}x_n$$

$$\cdots \cdots \cdots \cdots \cdots \cdots \tag{1.4}$$

$$\cdots \cdots \cdots \cdots \cdots \cdots$$

$$\frac{dx_n}{dt} = a_{n1}x_1 + a_{n2}x_2 \cdots a_{nn}x_n$$

Here the coefficients $a_{ij}(i,j=1,2,\ldots,n)$ form a set of n^2 real numbers, and the unknowns $x_j = x_j(t)$ are n real-valued functions of a real variable t .

A solution of (1.4) is conceived as a set of n real-valued differentiable functions, $x_i(t)$, $i = 1,2,\ldots,n$, verifying (1.4) for any $t \in R$ (the real line).

A better understanding of the system (1.4) is obtained through the introduction of the cartesian space (real) n-dimensional R^n ; the elements of R^n are "points" $x = (x_1, x_2 \cdots x_n)$, where x_i is the ith coordinate of the point x .

Consequently, a solution of (1.4) will be a curve in R^n : $x(t) = (x_1(t), x_2(t), \ldots x_n(t))$, that is a (continuous) map $x : R \to R^n$.   If every function $x_i(t)$ is differentiable, the map x is also (by definition)

2

differentiable; its derivative is defined by the equality

$\frac{dx}{dt} = x'(t) = (x'_1(t), x'_2(t), \ldots x'_n(t))$; therefore the derivative is again a

map of R into R^n .

One can express this derivative in the form:

$$x'(t) = \lim_{h \to 0} \frac{1}{h} (x(t + h) - x(t))$$

where the convergence is intended in the euclidean norm of R^n ,

$$\|x\| = [x_1^2 + x_2^2 + \ldots x_n^2]^{\frac{1}{2}} .$$

We can write (1.4) in an abbreviated form through the introduction of the

matrix

$$A = [a_{ij}] = \begin{bmatrix} a_{11} & a_{12} & \cdots & a_{1n} \\ a_{21} & a_{22} & \cdots & a_{2n} \\ \cdot & \cdot & \cdot \cdot \cdot \cdot \cdot & \cdot \\ a_{n1} & a_{n2} & \cdots & a_{nn} \end{bmatrix}$$

and afterwards of the associated map $R^n \to R^n$ defined by:

$$x = (x_1 \ldots x_n) \in R^n \to Ax = y \in R^n , \quad \text{where} \quad y_i = a_{i1}x_1 + \ldots + a_{in}x_n .$$

Thus, one can write (1.4) under the form

$$x'(t) = Ax(t) . \tag{1.5}$$

It is possible to obtain all the solutions of (1.5) by means of the formula

$$x(t) = \exp(tA)\cdot K , \quad \text{where} \quad K = (k_1 \ldots k_n) \in R^n \tag{1.6}$$

and the operator $\exp(tA)$ is defined by means of the series

$$\exp(tA) = I + \frac{t}{1!} A + \frac{t^2}{2!} A^2 + \ldots + \frac{t^n}{n!} A^n + \ldots \tag{1.7}$$

which is convergent in the space $L(R^n ; R^n)$ - of all linear maps, $R^n \to R^n$, with the uniform norm:

$$\|T\|_{L(R^n, R^n)} = \sup_{\|x\|_{R^n} \leq 1} \|Tx\|_{R^n} , \quad \forall T \in L(R^n, R^n) . \tag{1.8}$$

We define a Cauchy problem for (1.5) when we add an "initial" condition

$$x(t_o) = x_o \quad \text{given in} \quad R^n , \quad \text{for a fixed real number} \quad t_o . \tag{1.9}$$

We solve uniquely the equation (1.5) with the condition (1.9) by means of the formula

$$x(t) = \exp[(t - t_o)A] \cdot x_o . \tag{1.10}$$

Let us take the special case: $t_o = 0$, and let us denote: $\exp(tA) = U(t)$. This will be in fact a <u>family of linear operators</u>: $R^n \to R^n$, depending on the real <u>parameter t which runs over the real line.</u>   If one has a solution $x(t)$ of the equation (in R^n) ,

$$x'(t) = Ax(t) , \quad x(0) = x_o \in R^n \tag{1.11}$$

then $x(t) = U(t)x_o$ (at $t = 8$) and one can say that the family $\underline{U(t)}$ carries the <u>initial values of the solution in the values at the "instant" t .</u>

<u>Remark</u>.   This preliminary part is inspired by Ch. 1, §1 and 2, in [17]

§2.   <u>THE BANACH SPACE CASE</u>

We start from now on the investigation of the differential equation

$$u'(t) = Au(t) \tag{2.1}$$

4

in spaces which are more general than the euclidean but we still restrict
ourselves to Banach or Hilbert spaces, leaving aside more general (locally
convex) linear spaces.

Let us remember that a Banach space E is a vector space on the field of
real or complex numbers, a norm function $x \to \|x\|$ from E into R^+ being
defined on E with the usual properties, and any Cauchy sequence $(x_n)_1^\infty \subset E$
with respect to the norm $\|\ \|$ being convergent to an element $x \in E$.

Let us take in E a vector subspace D and afterwards a map A , defined
on D , with range into E . Accordingly

$$E \supset D \xrightarrow{A} R_A \subset E \quad \text{where} \quad R_A = \{x \in E, \ x = Ay \ \text{with} \ y \in D\} \ .$$

D is the domain of $A : D = D(A)$, while R_A is the range of A .

We are only concerned about *linear* operators; we assume therefore the
relations

$$\begin{aligned}
&A(x + y) = A(x) + A(y) \ , \quad x, y \in D(A) \\
&A(\lambda x) = \lambda A(x) \ , \forall \ x \in E \ , \forall \ \lambda \in R \quad (\text{or} \ \lambda \in C) \ .
\end{aligned} \tag{2.2}$$

We then envisage functions $x(t)$ defined over a finite or infinite interval
of the real line, having range in E ; that is to say: $R \supset (a,b) \xrightarrow{x(t)} E$.

Such a function is continuous in the point $t_o \in (a,b)$ if for any positive
number ε there exists a positive number $\delta(\varepsilon)$ in such a way that for
all $t \in (a,b)$ with $|t - t_o| < \delta$ it results $\|x(t) - x(t_o)\| < \varepsilon$; it
admits a right derivative (resp. left) for $t = t_o$ if there exists $y \in E$
such that: $\|\frac{1}{h} [x(t_o + h) - x(t_o)] - y\| \to 0$ for $h \downarrow 0$ (resp. $h \uparrow 0$) .

One writes in that case: $(\frac{d+}{dt} x)(t_o) = y$ (resp. $(\frac{d-}{dt} x)(t_o) = y)$. If both
right and left derivatives exist and coincide one says that $x(t)$ is
differentiable for $t = t_o$, its derivative being $x'(t_o) = (\frac{d}{dt} x)(t_o) = y$.

Of course, one can enlarge the concepts of continuity and differentiability, considering instead of one point t_o a whole interval (finite or infinite) of R^1 . If $x(t)$ is differentiable for any $t \in (a,b)$, its derivative $x'(t)$ will be again a function $(a,b) \to E$ and $x(t)$ will be continuous in (a,b) .

With this (minimal) set of definitions, we are able by now to describe a general Cauchy problem, in the following way. We take a function $u(t)$, $0 \leq t < \infty \to E$, which is differentiable for any $t > 0$ and has a right derivative for $t = 0$, such that $u(t) \in D(A)$ for all $t \in [0,\infty)$.

After choosing an element $u_o \in D(A)$, we seek for a solution $u(t)$ of the equation

$$u'(t) = Au(t) \ , \ \ t > 0 \ , \ \ (\tfrac{d+}{dt} u)(0) = Au(0) \ , \ \ u(0) = u_o \tag{2.3}$$

and we give the following

<u>Definition</u> *The Cauchy problem* (2.3) *is well-posed on the interval* $[0,\infty)$ *if, for every element* $u_o \in D(A)$ *there exists one and only one solution; and if, given any sequence* $(u_o^n)_1^\infty \subset D(A)$ *, such that* $u_o^n \to \Theta$ *when* $n \to \infty$ *,[1] the sequence of solutions* $u_n(t)$ *such that* $u_n(0) = u_o^n$ *converges to* Θ *for any real value of* t *.*

Let us consider then a well-posed Cauchy problem; one is able to associate to this problem a family of maps $U(t) : D(A) \to D(A)$, depending on the non-negative parameter t , defined by means of the formula: $U(t)u_o = u(t)$, $u(t)$ being the solution of (2.3).

One sees that $U(0) = I$ (the identity map) of $D(A)$; besides, $\lim_{t \downarrow 0} U(t)x = x$ for any $x \in D(A)$ and moreover $\lim_{\tau \to 0} \frac{1}{\tau} [U(\tau)x - x]$ exists and equals Ax , $\forall \ x \in D(A)$. With respect to the family of maps $U(t)$,

6

$D(A) \to D(A)$ it is possible to prove without difficulty some simple results. We start with

<u>Proposition 2.1</u> *The mapping $U(t)$ of $D(A)$ into itself is a linear transformation, for any $t \geq 0$.*

Let us take in fact two elements u_1, u_2 in $D(A)$; their sum, $u_1 + u_2$ belongs also to $D(A)$, and $U(t)(u_1 + u_2)$ is the solution of (2.3) which has $u_1 + u_2$ as "initial value". Let us consider also the functions: $u_1(t) = U(t)u_1$, $u_2(t) = U(t)u_2$, which have as "initial value" u_1 and u_2 respectively.

Their sum: $u_1(t) + u_2(t)$ is again a solution of (2.3) with initial value $u_1 + u_2$, just as the function $U(t)(u_1 + u_2)$. Using the uniqueness of the Cauchy problem, it follows that $U(t)(u_1+u_2) = U(t)u_1+U(t)u_2$, $t \geq 0$, $\forall u_1, u_2 \in D(A)$. In the same way one proves that $U(t)(\lambda x) = \lambda U(t)x$, $\forall \lambda$ real or complex (precisely $\forall \lambda$ in the field of scalars for E) , $\forall x \in D(A)$ and $\forall t \geq 0$. We have also

<u>Proposition 2.2</u> *For any sequence $(x_n)_1^\infty \subset D(A)$ such that: $\lim\limits_{n \to \infty} x_n = x_o \in D(A)$, $\lim\limits_{n \to \infty} U(t)x_n = U(t)x_o$, $t \geq 0$.*

This result is an immediate consequence of the fact that the above considered Cauchy problem is well posed; in fact $\lim\limits_{n \to \infty}(x_n - x_o) = \theta$; it follows therefore that: $\lim\limits_{n \to \infty} U(t)(x_n - x_o) = \theta$ also $\forall t \geq 0$.

A fundamental property in this context is the "semi-group" property of the family $U(t)$; precisely, $U(t)$ is a representation of the semi-group of real non-negative numbers into the set of all linear mappings of $D(A)$ into $D(A)$; that is to say, one has the following

---

[1] From now on "θ" represents the null element in the Banach space under consideration.

<u>Proposition 2.3</u>   *For any pair of real non-negative numbers* t_1, t_2 ,

$$U(t_1 + t_2)x = U(t_1)U(t_2)x = U(t_2)U(t_1)x , \quad \forall x \in D(A) .$$

Let in fact $u(t) = U(t)x$;   then consider, for a fixed positive number τ , the translated function $w_\tau(t)$, $t \geq 0 \to E$, which is defined by $w_\tau(t) = u(t + \tau)$;   one sees that $w_\tau'(t) = u'(t + \tau) = (Au)(t + \tau) = Aw_\tau(t)$ (this equality is true because A is an operator independent of t) .   One has also: $w_\tau(0) = u(\tau) = U(\tau)x$ and $w_\tau(t) = U(t + \tau)x$.   Let us consider also the function $w_1(t) = U(t)(U(\tau)x)$, which is the solution of the equation $w_1'(t) = Aw_1(t)$, verifying the condition $w_1(0) = U(\tau)x$. Therefore, it coincides with $w_\tau(t)$ which demonstrates the required result.

Let us remember now two simple facts of a more general nature:

<u>Lemma 2.1</u>   *Let* E *be a Banach space,* T *a linear operator of* $D(T) \subset E$ *into* E ,   *where* $D(T)$ *is a linear subset of* E . *Let us assume that* $\lim\limits_{n \to \infty} Tx_n = \theta$ *for every sequence* $(x_n)_1^\infty \subset D(T)$, *such that* $\lim\limits_{n \to \infty} \|x_n\| = 0$. *Then, there exists a positive constant* β, *such that* $\|Tx\| \leq \beta \|x\|, \forall x \in D(T)$ (see [37] page 43, Corollary 2).

<u>Lemma 2.2</u>   *In the assumptions of Lemma 2.1, there exists one and only one linear continuous transformation of* E *into* E , $\tilde{T}$, *such that* $\tilde{T}x = Tx$ $\forall x \in D(T)$, *provided that* $D(T)$ *is also dense in* E .
(see [13] -Th. II.2.1, page 50]) .

One can now give the following

<u>Proposition 2.4</u>   *Let us assume the domain* $D(A)$ *to be dense in* E ; *there exists then one and only one mapping* $V(t)$, *from* $t \geq 0$ *into* $L(E,E)$, *such that* $V(t)x = U(t)x$, $\forall x \in D(A)$ *and also such that* $V(t + s) = V(t)V(s) = V(s)V(t)$, $\forall s, t \in R^+$.

In fact, one can apply $\forall t \geq 0$ Lemma 2.2 taking $V(t) = \tilde{U}(t)$.   The relation $U(t)U(s)x = U(t + s)x$, $\forall x \in D(A)$ extends then by continuity to each element $x \in E$.

One calls $V(t)$ the semi-group of linear transformations associated to the well-posed Cauchy problem (2.3). In a more precise way, the mapping $V(t)$ of E into E is defined as follows:

For any $x \in E$, there exists a sequence $x_n \in D(A)$, $x_n \to x$. If $t \geq 0$ is fixed, the sequence $\{U(t)x_n\}_{n=1}^{\infty}$ is a Cauchy sequence, because an estimate: $\| U(t)x \| \leq \beta(t) \| x \|$, $\forall x \in D(A)$ is satisfied, according to Lemma 2.1. One puts therefore, as usual, $V(t)x = \lim_{n \to \infty} U(t)x_n$.

This limit is independent of the sequence $x_n \to x$, as one sees for example by the consideration of the alternate sequence $x_1 , x_1' , x_2 , x_2' , \cdots$ formed starting with another sequence $x_n' \to x$.

The semi-group property: $V(t_1 + t_2) = V(t_1)V(t_2)$ is proved then through a passage to the limit: if $x \in E$ and $x_n \in D(A)$, and if $x_n \to x$, one has: $V(t_1 + t_2)x = \lim_{n \to \infty} U(t_1 + t_2)x_n = \lim_{n \to \infty} U(t_1)U(t_2)x_n$. But: $x_n \in D(A)$ and $x_n \to x$ implies $U(t_2)x_n \in D(A)$ and $U(t_2)x_n \to V(t_2)x$.

Consequently: $\lim_{n \to \infty} U(t_1)(U(t_2)x_n) = V(t_1)V(t_2)x = V(t_1 + t_2)x$, $\forall x \in E$.

§3. STRONG CONTINUITY OF THE ASSOCIATED SEMI-GROUP

Using the equality $V(t)x = U(t)x = u(t)$, $\forall x \in D(A)$, one sees immediately that $V(t)x$ is continuous, $t \geq 0 \to E$, for any $x \in D(A)$. We shall demonstrate the

<u>Main Lemma</u> *For any number* $\delta > 0$, *there exists a constant* $M_\delta > 0$, *in such a way that one has* $\| V(t) \|_{L(E,E)} \leq M_\delta$, *if* $\delta \leq t \leq \frac{1}{\delta}$ (see [16] -page. 304]) .

But let us assume the Main Lemma already demonstrated; let us take now an arbitrary element $x \in E$, and let be $\delta \leq t \leq \frac{1}{\delta}$. There exists a sequence $(x_n)_1^\infty \subset D(A)$, such that $\lim_{n \to \infty} x_n = x$. It follows that:

$$\| V(t)x - V(t)x_n \| \leq M_\delta \| x - x_n \| \quad \text{for} \quad \delta \leq t \leq \frac{1}{\delta} \quad \text{and} \quad \lim_{n \to \infty} V(t)x_n = V(t)x \text{ takes}$$

place uniformly on the interval $[\delta, \frac{1}{\delta}]$; the functions $V(t)x_n$ being continuous, the function $V(t)x$ is also continuous on $[\delta, \frac{1}{\delta}]$. Now, any number $t > 0$ belongs to a convenient interval of this kind. One obtains therefore the

__Theorem 3.1__ *The function* $V(t)x$, $t > 0 \to E$ *is continuous for each element* $x \in E$.

One gives now the

Proof of the Main Lemma

Following the construction of the semi-group $V(t)$, if one fixes an element $x \in E$, one can choose a sequence $(x_n)_1^\infty \subset D(A)$, convergent to x, and it will follow that $V(t)x = \lim_{n \to \infty} U(t)x_n$, for all $t \geq 0$.

Consequently, the scalar function $t \to \| V(t)x \|$ is a pointwise limit of the functions: $t \to \| V(t)x_n \|$ which are continuous; accordingly, the function $t \to \| V(t)x \|$ is Lebesgue measurable, and hence the numerical sets, $\{t; \| V(t)x \| \leq \tilde{\alpha}\}$, are measurable for each real number $\tilde{\alpha}$.

On the other hand, if the Lemma were false, one could find a number $\delta_0 > 0$, and a sequence $(t_n)_1^\infty \subset [\delta_0, \frac{1}{\delta_0}]$ such that: $\| V(t_n) \|_{L(E,E)} \geq n$. One may even admit - after a convenient extraction - that the sequence $(t_n)_1^\infty$ is convergent: $\lim_{n \to \infty} t_n = \gamma \in [\delta_0, \frac{1}{\delta_0}]$. Using now the resonance theorem ([37] -page 69]) one infers the existence of an element $x_0 \in E$, such that the sequence $\| V(t_n)x_0 \|$ is unbounded. In that case, for a subsequence $(t_{n_q})_{q=1}^\infty \subset (t_n)_1^\infty$, it will even follow:

$$\| V(t_{n_q})x_0 \| \geq q, \quad q = 1,2,\ldots \quad \text{and} \quad \lim_{q \to \infty} t_{n_q} = \gamma \text{ too.}$$

10

Let us remark now that one may find a Lebesgue measurable set $F \subset [0,\gamma]$ such that $\text{meas } F > \frac{\gamma}{2}$ and $\sup\limits_{t \epsilon F} \| V(t)x_o \| = L < \infty$. Let in fact:

$$G_p = \{ t \; \epsilon \; [0,\gamma] \; , \; \| V(t)x_o \| \leq p \} \; , \quad p = 1,2,\ldots \; . \quad \text{The sets} \quad G_p \; \text{are}$$

measurable; one has $G_1 \subset G_2 \ldots$ and $\bigcup\limits_{p=1}^{\infty} G_p = [0,\gamma]$, in view of:

$\| V(t)x \| < \infty \; \forall t \geq 0$. Consequently: $\lim\limits_{p \to \infty} \text{meas } G_p = \gamma$ and therefore

$\text{meas } G_p > \frac{\gamma}{2}$ for p sufficiently large, for example if $p = p_o$; let us

put $F = G_{p_o}$; one has $\text{meas } F > \frac{\gamma}{2}$ and $\| V(t)x_o \| \leq p_o$ if $t \; \epsilon \; F$. Let

us also remark that $\text{meas } (F \cap [0,t_{n_q}]) > \frac{\gamma}{2}$ for sufficiently large q .

This is obvious for $t_{n_q} > \gamma$, when $F \cap [0,t_{n_q}] = F$; whereas, for

$t_{n_q} < \gamma$, $F = (F \cap [0,t_{n_q}]) \cup (F \cap (t_{n_q},\gamma])$, union of disjoint

subsets of F . Accordingly, $\text{meas } F = \text{meas}(F \cap [0,t_{n_q}]) + \text{meas}(F \cap (t_{n_q},\gamma]) > \frac{\gamma}{2}$

is equal to $\frac{\gamma}{2} + \varepsilon_o \; , \; \varepsilon_o > 0$.

But, if $q \geq q_o(\varepsilon_o)$, $\text{meas}(F \cap (t_{n_q},\gamma]) < \varepsilon_o$. Consequently:

$\varepsilon_o + \text{meas}(F \cap [0,t_{n_q}]) > \frac{\gamma}{2} + \varepsilon_o \; , \quad q \geq q_o$.

Let us consider then the translated set: $F_q = \{ t_{n_q} - \tau, \; \tau \; \epsilon \; F \cap [0,t_{n_q}] \}$;

Lebesgue's measure being invariant to translations, one sees that:

$\text{meas } F_q = \text{meas}(F \cap [0,t_{n_q}]) > \frac{\gamma}{2}$ for $q \geq q_o$. Let us take then a number

$\tau \; \epsilon \; F \cap [0,t_{n_q}]$; let us notice the semi-group relation: $V(t_{n_q})x_o =$

$V(t_{n_q} - \tau)V(\tau)x_o$ and afterwards the estimates: $q \leq \| V(t_{n_q})x_o \| \leq \| V(t_{n_q} - \tau) \|_{L(E,E)}$

$\| V(\tau)x_o \|_E \leq L \| V(t_{n_q} - \tau) \|_{L(E,E)}$, hence $\frac{q}{L} \leq \| V(\sigma) \|_{L(E,E)} \; , \; \forall \sigma \; \epsilon \; F_q$.

At this stage, one notices that the set: $\limsup\limits_{q \to \infty} F_q = \bigcap\limits_{n=1}^{\infty} \bigcup\limits_{q=n}^{\infty} F_q$ is non void.   In fact, let us put: $G_n = F_n \cup F_{n+1} \cup \ldots$.   Then:

$\lim\sup\limits_{q \to \infty} F_q = \bigcap\limits_1^\infty G_n$, where $G_1 \supset G_2 \supset \ldots G_n \supset \ldots$.   Besides,

meas G_1 = meas$\left(\bigcup\limits_1 G_n\right)$ is finite; in fact, if $\xi \in F_q$, $\xi = t_{n_q} - \tau$,

$0 \leq \xi \leq \dfrac{1}{\delta_0}$; hence: meas $G_1 \leq \dfrac{1}{\delta_0}$.   Consequently, meas $\bigcap\limits_1^\infty G_n = \lim\limits_{n \to \infty}$ meas G_n .

But $G_n \supset F_n$, $n = 1,2,\ldots$, hence: meas $G_n \geq$ meas $F_n > \dfrac{\gamma}{2}$ for n

sufficiently large.   In that case, meas $\bigcap\limits_1^\infty G_n \geq \dfrac{\gamma}{2} > 0$; the set

$\bigcap\limits_1^\infty G_n = \lim\sup\limits_{q \to \infty} F_q$ is non-empty (being of positive measure), and there exists

at least one number σ belonging to an infinity of sets F_q .

This would imply : $\sup\limits_{x \neq \theta} \dfrac{\| V(\sigma)x \|_E}{\| x \|_E} = \infty$, a contradiction which proves

the Main Lemma, and the Theorem too.

§4.   REGULARITY PROPERTIES

The results demonstrated above permit us to obtain certain supplementary

information on the solutions of a well-posed Cauchy problem.   One has

firstly the

Theorem 4.1   *Let be $u(t)$ a solution of the Cauchy problem (2.3).   Then, the strong derivative $u'(t)$ is continuous, $t > 0 \to E$.*

One has in fact that: $u(t) = U(t)u(o)$.   Now, the strong derivative

$u'(t) = \dfrac{d}{dt} U(t)u(o)$ equals the limit

$$\lim\limits_{\substack{h \to 0 \\ h > 0}} \frac{1}{h} [U(t+h)u(o) - U(t)u(o)] = \lim\limits_{h \downarrow 0} U(t)\frac{1}{h} [U(h)u(o) - u(o)] =$$

$$\lim\limits_{h \downarrow 0} U(t)[\frac{1}{h}(u(h) - u(o))] = \lim\limits_{h \downarrow 0} V(t) \frac{1}{h} (u(h) - u(o)) = V(t)u'_+(o)$$

$$= V(t)(Au(o)) .$$

The Theorem 4.1 is therefore a consequence of Theorem 3.1.

Remark 1   One has relations: $u'(t) = Au(t) = A[U(t)u(o)] = V(t)(Au(o))$.
But every element $x \in D(A)$ is equal to the initial value $u(o)$ of a certain
unique solution $u(t)$; it is also; $U(t)x = V(t)x$ if $x \in D(A)$; hence,
the equality $AV(t)x = V(t)Ax$ takes place for all $x \in D(A)$.

Let us put now the following

Definition 4.1   Let $\tilde{D} = \{x \in E , V(t)x$ possesses a strong right derivative
for $t = 0\}$ and let $\tilde{A}$ be defined from $\tilde{D}$ into E by
$$\tilde{A}x = \lim_{\eta \downarrow 0} \frac{1}{\eta} [V(\eta)x - x] .$$

One calls then $\tilde{A}$ the infinitesimal generator of the semi-group $V(t)$.
One has the following

Theorem 4.2   For any element $x \in \tilde{D}$, the function $V(t)x$ possesses a
strong derivative continuous for any $t > 0$.

One has firstly, if $h > 0$, the equality
$\frac{1}{h}[V(t + h)x - V(t)x] = V(t)[\frac{1}{h}(V(h)x - x)]$; the expression at right tends
obviously to $V(t)\tilde{A}x$.

Let us take also $h < 0$; one may write for $t > 0$ and h small enough,
the equality $\frac{1}{h}[V(t + h)x - V(t)x] = V(t + h)[-\frac{1}{h}(V(-h)x - x)]$ and one can
prove without difficulty that the limit at right is again $V(t)\tilde{A}x$.

The continuity of $\frac{d}{dt} V(t)x$ for $t > 0$ results then from Theorem 3.1.

Remark 2   If one takes $h > 0$ one has also the equality
$\frac{1}{h}[V(t + h)x - V(t)x] = \frac{1}{h}[V(h) - I]V(t)x$; if $x \in \tilde{D}$, the left-hand
expression tends to $V(t)\tilde{A}x$ if $h \to 0$; consequently $V(t)x \in \tilde{D}$ and
$V(t)\tilde{A}x = \tilde{A}V(t)x$; one can say that the semi-group associated to the well-

posed Cauchy problem commutes with its infinitesimal generator $\tilde{A}$.

<u>Remark 3</u>   One has that $A \subset \tilde{A}$, that is $\tilde{A}$ is an extension of A .
In fact, if $x \in D(A)$, $V(t)x = U(t)x = u(t)$, $u(0) = x$, and hence
$u'_+(0) = Au(0)$ exists; accordingly $x \in \tilde{D}$ and $Au(0) = \tilde{A}x$, $\forall x \in D(A)$.

2 Uniformly correct Cauchy problem and semi-groups of class C_o

<u>INTRODUCTION</u>

In this section we define, according to ([20,I,§2]) a Cauchy problem which is *uniformly* well posed; the associated semi-group is strongly continuous for any $t \geq 0$; its elementary properties are exposed in the usual way (see [7], [10], [11], [15], [16]) including the connection between the Laplace transform of the semi-group and the resolvent of its infinitesimal generator.

Afterwards one continues the study of the uniformly correct Cauchy problem including a treatment of the inhomogeneous equation, and of the well-posed Cauchy problem on the whole real line. We end this Chapter with the fundamental existence theorem (by Hille-Yosida-Phillips) and some of its "abstract" applications.

§1.   <u>THE UNIFORMLY WELL–POSED CAUCHY PROBLEM</u>

Let us start with the following simple

<u>Definition</u>: *A well-posed Cauchy problem for the equation $u'(t) = Au(t)$ is uniformly well-posed if $D(A)$ is dense in E and if given any sequence of solutions $\{u_n(t)\}_{n=1}^{\infty}$ such that $\lim_{n \to \infty} u_n(0) = \theta$, it results that $\lim_{n \to \infty} u_n(t) = \theta$, uniformly on any finite interval $0 \leq t \leq T$ where $T > 0$.*

For this class of Cauchy problems, more restrictive than the preceding, one has firstly the:

<u>Theorem 1.1</u>: *If $V(t)$ is the semi-group associated to an uniformly well-posed Cauchy problem, one has:* $\lim_{t \downarrow 0} V(t)x = x$, $\forall x \in E$. *(This means*

strong continuity of the semi-group $V(t)$ for $t \geq 0$).

One proves in fact that there is a $M_T > 0$, such that

$$\| V(t) \|_{L(E,E)} \leq M_T, \quad 0 \leq t \leq T, \quad \text{as a consequence of the inequality}$$

$$\| V(t)x \|_E = \| U(t)x \|_E \leq M_T \| x \|_E, \quad \forall x \in D(A), \quad \forall t \in [0,T].$$

But, if this last inequality would be false, one could find a sequence $(x_n)_1^\infty \subset D(A)$ and a sequence $(t_n)_1^\infty \subset [0,T]$ in such a way that $\| U(t_n)x_n \| \geq n \| x_n \|$, $n = 1,2,\ldots$.

Let us put now: $v_n = \dfrac{1}{n \| x_n \|} \cdot x_n$; one has $\| U(t_n)v_n \| \geq 1$, $n = 1,2,\ldots$ and also $\| v_n \| = \dfrac{1}{n} \to 0$ if $n \to \infty$.

If $u_n(t)$ is the solution to the Cauchy problem such that $u_n(0) = v_n$, $n = 1,2,\ldots$, one should have: $\lim\limits_{n \to \infty} u_n(t) = \theta$, uniformly for $0 \leq t \leq T$. But $u_n(t) = U(t)v_n$; consequently $\| u_n(t_n) \| \geq 1$, $\forall n = 1,2,\ldots$ absurd.

Finally, since $D(A)$ is dense in E and **because** $\lim\limits_{t \downarrow 0} V(t)x = x$, $\forall x \in D(A)$, one deduces the wanted result.

Consequently, the semi-group $V(t)$ is strongly continuous for all $t \geq 0$. This class of semi-groups is named C_o (see [16 - pag. 321]) and it possesses certain fundamental properties that we shall give below.

<u>Remark</u>: The equality (proved in Ch. I): $u'(t) = Au(t) = AV(t)u(0) = V(t)(Au(0))$ implies here the continuity of the derivative $u'(t)$ for $0 \leq t < \infty$.

§2. <u>FUNDAMENTAL PROPERTIES OF SEMI-GROUPS OF CLASS C_o</u>

The semi-groups $V(t)$ of class C_o satisfy an exponential estimate on the half-line $t \geq 0$ which is found very useful in successive developments.

One has precisely the [(*)]

Theorem 2.1: *For any semi-group $V(t)$ of class C_0 one can find two real constants, $M > 0$ and $\omega \geq 0$, in such a way that the inequality*

$$\| V(t) \|_{L(E,E)} \leq M \exp(\omega t) , \quad 0 \leq t < \infty \tag{2.1}$$

is verified.

Let us remark that one has: $\sup_{0 < t \leq 1} \| V(t)x \| = C_x < \infty$, $\forall x \in E$. According to the uniform boundedness theorem ([37 - pag. 68]), one deduces that, for a constant $M > 0$, holds: $\| V(t) \| \leq M$, $0 \leq t \leq 1$. Now, every $t > 0$ can be written in the form: $t = [t] + \{t\} = n + \tau$, where $0 \leq \tau < 1$. Consequently one has: $V(t) = V(n+\tau) = V(n)V(\tau) = [V(1)]^n V(\tau)$ and:

$$\| V(t) \| \leq \| V(1) \|^n \| V(\tau) \| \leq M \| V(1) \|^n = M \exp(n \log \| V(1) \|), \quad t \geq 0 .$$

If $\log \| V(1) \| \leq 0$, then $\| V(t) \| \leq M$, and the required estimate is obtained with $\omega = 0$. If on the other hand, one has $\log \| V(1) \| > 0$, then one sees that: $\| V(t) \| \leq M \exp(n \log \| V(1) \|) \exp(\tau \log \| V(1) \|) = M \exp(t \log \| V(1) \|) = M \exp(t\omega)$, where $\omega = \log \| V(1) \| > 0$.

We shall use now definition and first properties of the infinitesimal generator (Ch. I, §4), and prove the following

Theorem 2.2: *The domain $\tilde{D}$ of the infinitesimal generator $\tilde{A}$ for an arbitrary semi-group $V(t)$ of class C_0 is a subset linear and dense in E.*

In fact, the linearity being obvious, and since: $\lim_{t \downarrow 0} \dfrac{1}{t} \int_0^t V(\sigma)x d\sigma = x$, $\forall x \in E$, it will be sufficient to prove that, $\forall x \in E$ and $\forall t \geq 0$,

---

(*) A similar behaviour at ∞ for the more general semigroups considered in Chapter I can also be established (see [20, pag. 28-29]).

one has $\int_o^t V(\sigma)xd\sigma \varepsilon \tilde{D}$.   Let us remark for that purpose that one has:

$$\frac{1}{\eta} [V(\eta) - I] \int_o^t V(\sigma)xd\sigma = \frac{1}{\eta} \int_o^t [V(\eta + \sigma)x - V(\sigma)x]d\sigma =$$

$$\frac{1}{\eta} \int_\eta^{t+\eta} V(\sigma)xd\sigma - \frac{1}{\eta} \int_o^t V(\sigma)xd\sigma = - \frac{1}{\eta} \int_o^\eta V(\sigma)xd\sigma + \frac{1}{\eta} \int_t^{t+\eta} V(\sigma)xd\sigma \quad .$$

Now, if one lets η tend towards 0 , one obtains that the limits at the right exist and therefore: $\int_o^t V(\sigma)xd\sigma \varepsilon \tilde{D}$ and also: $\tilde{A} \int_o^t V(\sigma)xd\sigma = V(t)x{-}x$, $\forall x \varepsilon E$.

Remark: One has besides the equality: $\tilde{A} \int_o^t V(\sigma)xd\sigma = \int_o^t V(\sigma)\tilde{A}xd\sigma$, $\forall x \varepsilon \tilde{D}$, $\forall t \geq 0$.

In fact, if $x \varepsilon \tilde{D}$,

$$A \int_o^t V(\sigma)xd\sigma = \lim_{\eta \downarrow 0} \frac{1}{\eta} [V(\eta) - I] \int_o^t V(\sigma)xd\sigma$$

$$= \lim_{\eta \downarrow 0} \int_o^t V(\sigma) \frac{1}{\eta} [V(\eta) - I]xd\sigma = \int_o^t V(\sigma)\tilde{A}xd\sigma \quad .$$

We can prove now the

Theorem 2.3: *The infinitesimal generator $\tilde{A}$ of a semi-group $V(t)$ of class C_o is a closed operator* (i.e. it has a closed graph in $E \times E$) .

In fact, let us assume that $(x_n)_1^\infty \subset \tilde{D}$ and that $x_n \to x_o$ in E while $\tilde{A}x_n \to y_o$ in E . One has also the equality: $V(t)x - x = \tilde{A} \int_o^t V(\sigma)xd\sigma = \int_o^t V(\sigma)\tilde{A}xd\sigma$ (remark above) $(x \varepsilon \tilde{D})$. Consequently one obtains:

$$V(t)x_n - x_n = \int_o^t V(\sigma)\tilde{A}x_n d\sigma , \quad n = 1,2,\ldots .$$

For $n \to \infty$ one deduces without difficulty that: $V(t)x_o - x_o = \int_0^t V(\sigma)y_o d\sigma$; this entails that $x_o \in \tilde{D}$ and $\tilde{A}x_o = y_o$.

§3. <u>THE LAPLACE TRANSFORM OF THE SEMI-GROUPS OF CLASS C_o</u>

If $V(t)$ is a semi-group of class C_o , one may consider, for any real or complex number λ , the improper integral

$$\int_0^\infty e^{-\lambda t} V(t)x dt , \quad \text{where} \quad x \in E . \tag{3.1}$$

In view of the inequality (2.1) one has that integral (3.1) is convergent for $\mathrm{Re}\ \lambda > \omega$; if one puts, for such λ , $R(\lambda)x = \int_0^\infty e^{-\lambda t} V(t)x dt$, one obtains the inequality: $\|R(\lambda)x\| \leq \int_0^\infty \exp((-\mathrm{Re}\ \lambda)t)M \exp(\omega t)\|x\| dt$

$= \dfrac{M}{\mathrm{Re}\,\lambda - \omega} \|x\| , \quad \forall x \in E .$

The mapping $R(\lambda)$ thus defined in the half-plane $\mathrm{Re}\ \lambda > \omega$, is linear continuous from E into E . One has also that $R(\lambda)x \in \tilde{D}$, $\forall x \in E$, $\tilde{D}$ being the domain of the infinitesimal generator of $V(t)$.

In fact, one has for all $x \in E$ the equality:

$$\frac{1}{\eta}[V(\eta)-I]\int_0^\infty e^{-\lambda t}V(t)x dt = \frac{1}{\eta}\int_0^\infty e^{-\lambda t}[V(t+\eta)x - V(t)x]dt =$$

$$\frac{1}{\eta}\int_\eta^\infty e^{-\lambda(\sigma-\eta)}V(\sigma)x d\sigma - \frac{1}{\eta}\int_0^\infty e^{-\lambda\sigma}V(\sigma)x d\sigma = \frac{1}{\eta}(e^{\lambda\eta}-1)\int_0^\infty e^{-\lambda\sigma}V(\sigma)x d\sigma$$

$$-\frac{1}{\eta}\int_0^\eta e^{-\lambda(\sigma-\eta)}V(\sigma)x d\sigma = \frac{1}{\eta}(e^{\lambda\eta}-1)R(\lambda)x - e^{\lambda\eta}\frac{1}{\eta}\int_0^\eta e^{-\lambda\sigma}V(\sigma)x d\sigma .$$

The right-hand side expression has a limit for $\eta \downarrow 0$ (this limit equals $\lambda R(\lambda)x - x$) ; therefore, the Laplace transform $R(\lambda)x = \int_0^\infty e^{-\lambda t}V(t)x dt$ belongs to $\tilde{D}$ and one has: $\tilde{A}\,R(\lambda)x = \lambda R(\lambda)x - x$, $\forall x \in E$, or also

19

$$(\lambda I - \tilde{A})R(\lambda)x = x \,, \quad \forall x \, \varepsilon \, E \,.$$

Let us take now the element $\quad x \quad$ in $\tilde{D}$. One has then

$$\frac{1}{\eta}\,[V(\eta) - I]R(\lambda)x = \int_0^\infty e^{-\lambda t}\,\frac{1}{\eta}\,[V(t+\eta) - V(t)]x\,dt = \int_0^\infty e^{-\lambda t}V(t)\frac{1}{\eta}[V(\eta)-I]x\,dt.$$

If $\quad \eta \downarrow 0 \quad$ one obtains hence: $\quad \tilde{A}\,R(\lambda)x = \int_0^\infty e^{-\lambda t}\,V(t)\tilde{A}x\,dt = R(\lambda)\tilde{A}x$.

Consequently one deduces the equality: $\quad R(\lambda)(\lambda I - \tilde{A})x = x \,, \quad \forall x \, \varepsilon \, \tilde{D}$. Therefore we have just proved the

<u>Theorem 3.1</u>: *If $V(t)$ is a semi-group of class C_o , verifying the estimate: $\| V(t) \| \leq M \exp(\omega t)$, $0 \leq t < \infty$, the operator $R(\lambda)$ defined by* (3.1) *for* Re $\lambda > \omega$ *is the resolvent operator $R(\lambda;\tilde{A})$ of the infinitesimal generator $\tilde{A}$ of the semi-group $V(t)$.*

§4. <u>THE UNIFORMLY WELL–POSED CAUCHY PROBLEM–SEQUEL</u>

Let us return to a uniformly well-posed Cauchy problem for the equation $u' - Au = \theta$, and let be $V(t)$ the associated C_o-semi-group. If $\tilde{A}$ is the generator of $V(t)$, it was already seen even in the more general case of (simple) well-posed Cauchy problem (Ch. I, §4), that the inclusion $A \subset \tilde{A}$ takes place.

In the present case one obtains the following more precise result:

<u>Theorem 4.1</u>: *The generator $\tilde{A}$ is the closure of the operator A .*

One has proved already that $\tilde{A}$ is closed. Because of: $A \subset \tilde{A}$ it follows that $\bar{A}$ = closure of A is also contained in $\tilde{A}$: $\bar{A} \subset \tilde{A}$. Afterwards we remark the relation

$$\int_0^\infty e^{-\lambda t}\,V(t)x\,dt \; \varepsilon \; D(\bar{A}) \quad \text{for any} \quad x \, \varepsilon \, D(A) \,. \tag{4.1}$$

In fact, if $V(t)x = u(t)$ and $x \, \varepsilon \, D(A)$, one has $x = u(0)$ and

20

$u'(t) = Au(t)$, $0 \leq t < \infty$. One has therefore: $\|u(t)\| \leq M \exp(\omega t)\|u(0)\|$ and the Laplace integral: $\int_0^\infty e^{-\lambda t} u(t)dt$ is well-defined for $\mathrm{Re}\ \lambda > \omega$. If one considers $N > 0$ and one carries out an integration by parts, one deduces the relation

$$\int_0^N e^{-\lambda t} u(t)dt = -\frac{1}{\lambda} e^{-\lambda t} u(t)\Big|_0^N + \frac{1}{\lambda} \int_0^N e^{-\lambda t} Au(t)dt \ .$$

When $N \to \infty$ one deduces for $\mathrm{Re}\ \lambda > \omega$ the existence of the improper integral: $\int_0^\infty e^{-\lambda t} Au(t)dt$ and the equality

$$\int_0^\infty e^{-\lambda t} u(t)dt = \frac{1}{\lambda} u(0) + \frac{1}{\lambda} \int_0^\infty e^{-\lambda t} Au(t)dt = \frac{1}{\lambda} u(0) + \frac{1}{\lambda} \int_0^\infty e^{-\lambda t} \bar{A}u(t)dt$$

(because $A \subset \bar{A}$) . Hence, the improper integral: $\int_0^\infty e^{-\lambda t} u(t)dt$ is convergent at the same time as the improper integral: $\int_0^\infty e^{-\lambda t} \bar{A}u(t)dt$. Therefore (see for example [21, Th. 1.3.5]) $\int_0^\infty e^{-\lambda t} u(t)dt \ \varepsilon \ D(\bar{A})$, i.e. (4.1), and we obtain also the equality

$$\bar{A} \int_0^\infty e^{-\lambda t} u(t)dt = \int_0^\infty e^{-\lambda t} \bar{A}u(t)dt = \lambda \int_0^\infty e^{-\lambda t} u(t)dt - u(0)$$

or

$$(\lambda I - \bar{A}) \int_0^\infty e^{-\lambda t} u(t)dt = u(0) = (\lambda I - \bar{A}) \int_0^\infty e^{-\lambda t} V(t)u(0)dt$$

and therefore: $(\lambda I - \bar{A})R(\lambda;\tilde{A})x = x$, $\forall x \ \varepsilon \ D(A)$.

Now, let be x_o arbitrarily choosen in E and let $(x_n)_1^\infty \subset D(A)$, $x_n \to x_o$; one deduces: $R(\lambda;\tilde{A})x_n \to R(\lambda;\tilde{A})x_o$ and also $(\lambda I - \bar{A})R(\lambda;\tilde{A})x_n = x_n \to x_o$. Because of closedness of $\lambda I - \bar{A}$ it follows that: $R(\lambda;\tilde{A})x_o \ \varepsilon \ D(\bar{A})$ and that $(\lambda I - \bar{A})R(\lambda;\tilde{A})x_o = x_o$, $\forall x_o \ \varepsilon \ E$.

Hence $D(\tilde{A}) = R(\lambda;\tilde{A})(E)$ is contained in $D(\bar{A})$ and we obtain the equality $D(\tilde{A}) = D(\bar{A})$ so that the inclusion $\bar{A} \subset \tilde{A}$ becomes the equality $\bar{A} = \tilde{A}$.

We shall end this section with

<u>Theorem 4.2</u>: *If $W(t)$ is a C_o-semi-group and if B is its infinitesimal generator, the Cauchy problem*

$$w'(t) = Bw(t) \, , \quad w(0) = w_o \in D(B) \tag{4.2}$$

is uniformly well-posed.

In fact, a solution to problem (4.2) on the interval $0 \le t < \infty$ is given by the formula: $w(t) = W(t)w_o$: by virtue of the estimate: $\| W(t) \| \le Me^{\omega t}$, one deduces the continuous dependence on the initial data, which is uniform on each compact subinterval of $[0,\infty)$.

It remains to demonstrate the uniqueness of the Cauchy problem (4.2): Let hence $z(t)$ be another solution of (4.2), such that $z(0) = w_o$. Now, for a fixed $t \ge 0$, one considers the function: $F_t(s) = W(t-s)(z(s)-w(s))$ $= W(t-s)u(s)$, obviously defined for $0 \le s \le t$, where $u(0) = \theta$; besides, $F_t(o) = W(t)(z(0) - w(0)) = \theta$ and also $u'(s) = Bu(s)$. Let us take now $0 < s < t$; the function $F_t(s)$ is (strongly) differentiable, and its strong derivative equals θ . This is obtained in the following way: One has firstly, for small enough τ , the equality:

$$\frac{1}{\tau} [F_t(s + \tau) - F_t(s)] = \frac{1}{\tau} [W(t - s - \tau)u(s + \tau) - W(t - s)u(s)] =$$

$$\frac{1}{\tau} [W(t - s - \tau) - W(t - s)]u(s) + W(t - s - \tau)[\frac{1}{\tau}(u(s + \tau) - u(s))] \tag{4.3}$$

Let be $\tau > 0$ and sufficiently small; (4.3) is written then as

$-W(t - s - \tau)\frac{1}{\tau} [W(\tau) - I]u(s) + W(t - s - \tau)\frac{1}{\tau}[u(s + \tau) - u(s)]$, while for

$\tau < 0$ it equals

$$-W(t-s)\left[\frac{W(-\tau) - I}{-\tau}\right] u(s) + W(t - s - \tau)\frac{1}{\tau}\left[u(s + \tau) - u(s)\right] \ .$$

In view of: $u(s) \in D(B)$, one has: $\lim\limits_{\substack{\tau>0\\\tau\to 0}} \dfrac{W(\tau) - I}{\tau} u(s) = Bu(s)$.

Likewise, $\dfrac{1}{\tau}\left[u(s + \tau) - u(s)\right] \to u'(s)$; now, one remarks equality

$$W(t-s-\tau)\frac{1}{\tau}\left[u(s+\tau) - u(s)\right] - W(t-s)u'(s) = W(t-s-\tau)\{\frac{1}{\tau}(u(s+\tau) - u(s)) - u'(s))\}$$

$$+ \left[W(t-s-\tau) - W(t-s)\right]u'(s)$$

and as a consequence, the estimate

$$\left\| W(t-s-\tau)(\frac{1}{\tau}(u(s+\tau) - u(s)) - W(t-s)u'(s)\right\| \le$$

$$Me^{\omega(t-s-\tau)}\left\|\frac{1}{\tau}(u(s+\tau) - u(s)) - u'(s)\right\| \left[W(t-s-\tau) - W(t-s)\right]u'(s)\| \ .$$

Using also the strong continuity of the semi-group $W(t)$, one obtains:

$$\lim\limits_{\tau\to 0} W(t-s-\tau)\left[\frac{1}{\tau}(u(s+\tau) - u(s))\right] = W(t-s)u'(s) \ .$$

In the same way one can see that

$$\lim\limits_{\tau\downarrow 0} W(t-s-\tau)\frac{1}{\tau}\left[W(\tau) - I\right]u(s) = \lim\limits_{\tau\uparrow 0} W(t-s)\left[-\frac{1}{\tau}\{W(-\tau) - I\}\right]u(s)$$

$$= W(t-s)Bu(s) \ .$$

One obtains therefore, in the end, for $0 < s < t$, the equality

$$\frac{d}{ds}F_t(s) = - W(t-s)Bu(s) + W(t-s)u'(s) = \theta \ .$$

Consequently, by integration from ε to s, where $\varepsilon > 0$, one finds

equality $F_t(s) = F_t(\varepsilon)$, $\varepsilon \le s < t$. Let us remark now that $\lim\limits_{\varepsilon\downarrow 0} F_t(\varepsilon) =$

$\lim_{\varepsilon \downarrow 0} W(t-\varepsilon)u(\varepsilon) = \theta$ (in view of $u(0) = \theta$) . Then $F_t(s)$ which does not

depend on ε , is too $= \theta$. In that case: $\lim_{t \downarrow s} F_t(s) = \lim_{t \downarrow s} W(t-s)u(s) =$

$u(s) = \theta$, $\forall s > 0$, which proves the required result.

§5. <u>INTEGRATION OF THE NONHOMOGENEOUS EQUATION</u>

We consider here the Cauchy problem

$$w'(t) = Bw(t) + f(t) , \quad w(o) = w_o \in D(B) \tag{5.1}$$

which is called "nonhomogeneous" when $f(t) \not\equiv \theta$. One assumes, as in Theorem 4.2 , that B is the infinitesimal generator of a semi-group $W(t)$ of class C_o .

The uniqueness of the solutions follows from Theorem 4.2; some sufficient conditions for the existence of the solution are given in the following lines.

<u>Theorem 5.1</u>: *Let us assume that* $f(t), 0 \leq t < \infty \to E$ *is continuously differentiable. In that case for any element* $u_o \in D(B)$ *the function*

$$u(t) = W(t)u_o + \int_0^t W(t-s)f(s)ds \tag{5.2}$$

is a solution of the equation (5.1) *such that* $u(0) = u_o$.

It suffices in fact to prove that the function

$$g(t) = \int_0^t W(t-s)f(s)ds , \quad 0 \leq t < \infty \to E \tag{5.3}$$

is a solution of the equation (5.1) such that $g(0) = \theta$.

We shall see firstly that: $g_+(t) = \lim_{h \downarrow 0} \frac{1}{h} [g(t+h) - g(t)]$ exists for all $t \geq 0$; one has

$$g(t) = \int_0^t W(\sigma)f(t-\sigma)d\sigma \ , \quad \text{which gives for} \quad h > 0 \ , \quad \text{the equality}$$

$$g(t+h) = \int_0^{t+h} W(\sigma)f(t+h-\sigma)d\sigma = \int_0^t W(\sigma)f(t+h-\sigma)d\sigma + \int_t^{t+h} W(\sigma)f(t+h-\sigma)d\sigma$$

$$(5.4)$$

and therefore

$$\frac{1}{h}\left[g(t+h) - g(t)\right] = \frac{1}{h}\int_0^t W(\sigma)\left[f(t+h-\sigma) - f(t-\sigma)\right]d\sigma$$

$$+ \frac{1}{h}\int_t^{t+h} W(\sigma)f(t+h-\sigma)d\sigma \ . \tag{5.5}$$

One can see without difficulty that:

$$\lim_{h\downarrow 0} \int_0^t W(\sigma)\frac{1}{h}\left[f(t+h-\sigma) - f(t-\sigma)\right]d\sigma = \int_0^t W(\sigma)f'(t-\sigma)d\sigma$$

and that:

$$\lim_{h\downarrow 0} \frac{1}{h}\int_t^{t+h} W(\sigma)f(t+h-\sigma)d\sigma = W(t)f(o)$$

and one deduces that the right-derivative $g'_+(t)$ exists, is continuous, and equals: $W(t)f(o) + \int_0^t W(\sigma)f'(t-\sigma)d\sigma$.

We give now the

Lemma 5.1: *If the continuous function* $h(t)$ *,* $0 \leq t < \infty \to E$ *admits continuous right-derivative:* $h'_+(t) = k(t)$ *,* $0 \leq t < \infty \to E$ *, then* $h(t)$ *is strongly differentiable and* $h'(t) = k(t)$ *,* $0 < t < \infty$ *.*

Let in fact : $K(t) = \int_0^t k(\tau)d\tau$; then $K'(t) = k(t) = h'_+(t)$. The function $K-h$ admits right-derivative $= \theta$; but every function $\phi(t)$,

$0 \leq t < \infty \to E$ such that $\phi'_+(t) = \theta \, \forall t \geq 0$ is a constant function (equal to $\phi(0)$).

We apply in the usual way Hahn-Banach's theorem (see [16, Th. 3.2.1]) and reduce ourselves to the case of real-valued functions. Let hence $\psi(t)$, $0 \leq t < \infty \to R$ be a continuous function and assume that $\psi'_+(t) = 0$ $t>0$. Let us take a couple of numbers $0 \leq a < b < \infty$; if $\psi(a) \neq \psi(b)$, for example, $\psi(b) - \psi(a) < 0$ and there exists $\varepsilon > 0$, such that $\psi(b) - \psi(a) + \varepsilon(b-a) < 0$ too.

Consider next the function: $\varphi_\varepsilon(t) = \psi(t) - \psi(a) + \varepsilon(t-a)$; one has $\frac{d_+}{dt} \varphi_\varepsilon = \psi'_+(t) + \varepsilon = \varepsilon > 0$. Consequently: $\lim\limits_{\delta \downarrow 0} \frac{1}{\delta} [\varphi_\varepsilon(a+\delta) - \varphi_\varepsilon(a)] =$

$\lim\limits_{\delta \downarrow 0} \frac{1}{\delta} \varphi_\varepsilon(a+\delta) = \varepsilon > 0$ and therefore $\varphi_\varepsilon(a+\delta_o) > 0$ for a $\delta_o > 0$. On the other hand one sees that $\varphi_\varepsilon(b) < 0$. Thus, one finds a point $\xi \in (a+\delta_o, b)$ such that $\varphi_\varepsilon(\xi) = 0$ and $\varphi_\varepsilon(\xi) \leq 0$ for $\xi \leq t \leq b$. (See [37 - pag. 239-240].) In that case: $\frac{d_+}{dt} \varphi_\varepsilon \Big|_{t=\xi} = \lim\limits_{\eta \downarrow 0} \frac{1}{\eta} \varphi_\varepsilon(\xi+\eta) \leq 0$, a contradiction which proves Lemma 5.1.

Accordingly, the strong derivative $g'(t)$ exists and equals:
$$W(t)f(o) + \int_o^t W(\sigma)f'(t-\sigma)d\sigma , \quad \forall t > 0 , \quad \text{and} \quad g'_+(o) = f(o) .$$
On the other hand, for every $h > 0$

$$\frac{1}{h} [g(t+h) - g(t)] = \frac{1}{h} \{ \int_o^{t+h} W(t+h-s)f(s)ds - \int_o^t W(t-s)f(s)ds \} =$$

$$\int_o^t \frac{1}{h}[W(h) - I]W(t-s)f(s)ds + \frac{1}{h} \int_t^{t+h} W(t+h-s)f(s)ds =$$

$$\frac{1}{h} [W(h) - I] \int_o^t W(t-s)f(s)ds + \frac{1}{h} \int_t^{t+h} W(t+h-s)f(s)ds . \tag{5.6}$$

But, obviously: $\lim\limits_{h\downarrow 0} \frac{1}{h} \int\limits_{t}^{t+h} W(t+h-s)f(s)ds$ exists and equals $f(t)$

therefore

$$\lim\limits_{h\downarrow 0} \frac{1}{h} [W(h) - I] \left(\int\limits_{o}^{t} W(t-s)f(s)ds\right)$$

exists (and $= g'(t) - f(t)$).

By definition of infinitesimal generator, one can infer that $\int\limits_{o}^{t} W(t-s)f(s)ds \ \varepsilon \ D(B)$ and that $B(\int\limits_{o}^{t} W(t-s)f(s)ds) = g'(t) - f(t)$, so that $g'(t) = Bg(t) + f(t)$, which proves Theorem 5.1.

A similar result is expressed in

<u>Theorem 5.2</u>: *Let us assume that the function $f(t)$, $0 \leq t < \infty \to D(B)$ is E-strongly continuous and that $(Bf)(t)$ has the same property. In that case, the function $g(t) = \int\limits_{o}^{t} W(t-s)f(s)ds$ belongs to $D(B)$, is strongly differentiable (at the right for $t = 0$) and verifies the equality $g'(t) = Bg(t) + f(t)$.*

If one uses Remark 2 in Ch. I, Section 4, one finds that for fixed $\sigma \leq t$, the element $W(t-\sigma)f(\sigma) \ \varepsilon \ D(B)$ and $B(W(t-\sigma)f(\sigma)) = W(t-\sigma)(Bf(\sigma))$; besides, the function $W(t-\sigma)(Bf)(\sigma)$ is continuous as a function of σ on the interval $[0,t]$, and using Th. 1.3.5 in [21] one obtains that $\int\limits_{o}^{t} W(t-\sigma)f(\sigma)d\sigma \ \varepsilon \ D(B)$ and that $B(\int\limits_{o}^{t} W(t-\sigma)f(\sigma)d\sigma) = \int\limits_{o}^{t} W(t-\sigma)(Bf)(\sigma)d\sigma$. Let us use again (5.6):

we find that the right-derivative $g'_+(t)$ exists and equals $f(t) + B \int\limits_{o}^{t} W(t-\sigma)f(\sigma)d\sigma$. Then we apply Lemma 5.1 in order to complete the proof.

We end this section with a few more observations:

Remark 1: There is a simple example which points out that, if $f(t)$, $0 \le t < \infty \to E$ does not verify the above stated assumptions (say, is not differentiable or has range outside $D(B)$), then, for a semi-group $W(t)$ of class C_o the above considered integral: $g(t) = \int_o^t W(t-\sigma)f(\sigma)d\sigma$ is not necessarily continuously differentiable. Let us take in fact an element $x \in E$, such that $x \notin D(B)$, and then $f(s) = W(s)x$, a strongly continuous function. Then $g(t) = \int_o^t W(t-\sigma)W(\sigma)x\,d\sigma = \int_o^t W(t)x\,d\sigma = tW(t)x$. If g would be differentiable for $t > 0$, then $W(t)x = \frac{1}{t} g(t)$ would be too. That this is not always the case can be seen for some examples (the semi-group of translations on $C[0,\infty]$), or more generally, when $W(t)$ is a *group* of operators, rather then a semi-group (see next section, Prop. 2) (see [9] or also [29] - page 296).

Remark 2: The following considerations are certainly useful in the study of the nonhomogeneous equation (see [31] and [29], Lemma 4.2, pag. 297):

Let be A the infinitesimal generator of a semi-group $T(t)$ of class C_o, in the Banach space E; let $f(t)$ be a continuous function, $0 \le t < \infty \to E$ and $u_o \in D(A)$ be given. Consider again the function $u(t) = T(t)u_o + \int_o^t T(t-\sigma)f(\sigma)d\sigma$. We prove the following.

Proposition: *The function $u(t)$ is E - continuously differentiable for $0 \le t < \infty$ and is a solution to the equation $u'(t) = Au(t) + f(t)$ if and only if the function $v(t) = \int_o^t T(t-\sigma)f(\sigma)d\sigma$ belongs to $D(A)$ for all $t \ge 0$, and $(Av)(t)$ is continuous, $0 \le t < \infty \to E$.*

Proof of the necessity: Since $w(t) = T(t)u_o$ verifies $w' = Aw$, then $u - w = v$ is a solution of $v' = Au + f - Aw = Av + f$; $v \in C^1[o,\infty;E]$ as u and w, hence $Av = v' - f \in C[o,\infty;E]$, as required.

28

<u>Proof of the sufficiency</u>: It is enough to show that $v'(t)$ exists and equals $Av + f$ on $[o,\infty)$. Using again Lemma 5.1, we consider the right-derivative $v'_+(t)$. Let us take, $\forall h > 0$, the expression

$$\frac{1}{h}\,[v(t+h) - v(t)] = \frac{1}{h}\,\{\int_o^{t+h} T(t+h-\sigma)f(\sigma)d\sigma - \int_o^t T(t-\sigma)f(\sigma)d\sigma\} =$$

$$\frac{1}{h}\,[\int_o^t T(h)T(t-\sigma)f(\sigma)d\sigma - \int_o^t T(t-\sigma)f(\sigma)d\sigma + \int_t^{t+h} T(t+h-\sigma)f(\sigma)d\sigma] =$$

$$\frac{1}{h}[T(h) - I] \int_o^t T(t-\sigma)f(\sigma)d\sigma + \frac{1}{h} \int_t^{t+h} T(t+h-\sigma)f(\sigma)d\sigma \ .$$

When $h \downarrow 0$, the last integral converges to $f(t)$ while the first converges to $A \int_o^t T(t-\sigma)f(\sigma)d\sigma$.

Hence $v'(t)$ exists and $= Av(t) + f(t)$, a continuous function on $[o,\infty)$. Finally, we note the following

<u>Corollary</u>: *Assume* $f(t)$, *E-continuous on* $[o,\infty)$ *and* $v(t) = \int_o^t T(t-\sigma)f(\sigma)d\sigma = \theta \ \forall t \geq 0$. *Then* $f(t) = \theta \ \forall t \geq 0$.

In fact $v(t) = \theta \ \epsilon \ D(A)$, $Av = \theta$ is continuous; hence $v' = Av + f = \theta = f(t)$, $\forall t \geq 0$.

§6.   <u>WELL-POSED CAUCHY PROBLEM ON THE REAL LINE</u>

Let be again E a Banach space; then A is a linear operator in E with dense domain $D(A)$; let be $u(t)$ a function, $-\infty < t < \infty \to D(A)$, which has a strong derivative in E , $u'(t)$, $\forall t \ \epsilon \ R$. We assume that

$$u'(t) = Au(t) \ , \quad -\infty < t < \infty \tag{6.1}$$

$$u(0) = u_o \ \epsilon \ D(A) \tag{6.2}$$

We say that (6.1) - (6.2) is a well-posed problem if

(i)   for any $u_o \in D(A)$ there exists a unique solution

(ii)   if $x_n \in D(A)$ and $x_n \to \theta$ and if $u_n'(t) = Au_n(t)$, $u_n(0) = x_n$,

       then $u_n(t) \to \theta$ $\forall$ real t .

One may associate to this problem the family of mappings $U(t)$, $D(A) \to D(A)$ depending on the real parameter $t \in (-\infty,\infty)$, defined by the formula

$$U(t)u_o = u(t) \quad \text{where} \quad u(t) \quad \text{is the solution of (6.1) - (6.2).}$$

We see that $U(0) = I$ (the identity map) and that $U(t)x \to x$ $\forall$ $x \in D(A)$ if $t \to 0$; besides, $\lim_{\tau \to 0} \frac{1}{\tau} [U(\tau)x - x] = Ax$, $x \in D(A)$. One verifies without difficulty that

$$U(t) \quad \text{is a linear mapping,} \quad D(A) \to D(A) , \quad \forall t \in R \qquad (6.3)$$

$$U(t)x_n \to U(t)x_o \quad \text{if} \quad x_n \to x_o \in D(A) , \quad \forall t \in R \qquad (6.4)$$

$$U(t)x \quad \text{is E-strongly continuous with respect to} \quad t , \quad \forall x \in D(A) \qquad (6.5)$$

$$U(t_1 + t_2)x = U(t_1)U(t_2)x = U(t_2)U(t_1)x , \quad \forall t_1 , t_2 \in R , \quad \forall x \in D(A)$$
$$\qquad (6.6)$$

this last equality as a consequence of the uniqueness of Cauchy problem and of the independence of A on t . Afterwards, just as in Chapter I, one can extend by continuity, using density of $D(A)$ in E , the family of operators $U(t)$ to a group of operators, $V(t) \in L(E,E)$, $t \in R$, that is verifying (6.6) for all $x \in E$ and $V(0) = I$, the identity in E . The first basic result is obviously related to the "Main Lemma" in Ch. I, §3. We state in fact the following

__Theorem 6.1:__ *The group $V(t)$ verifies a uniform estimate:*

$$\| V(t) \|_{L(E,E)} \leq M_T \quad if \ -T \leq t \leq T, \quad for \ any \ T > 0 .$$

__Proof:__ One remarks firstly that the function $V(t)x$, $t \in R \to E$, is (strongly) measurable $\forall x \in E$, so that the scalar function: $t \in R \to \| V(t)x \|$ has the same property. Now, if the theorem were false, one could find a $T > 0$ such that $\| V(t) \|_{L(E,E)}$ is unbounded on $|t| \leq T$; hence, $\exists (t_n)_1^{\infty} \subset [-T,T]$, such that $\| V(t_n) \| \geq n$, $n = 1,2,\ldots$. Besides, from the uniform boundedness theorem, there exists an element $x_o \in E$ such that the sequence $V(t_n)x_o$ is unbounded; again one may assume (after one more extraction) that $\| V(t_n)x_o \| \geq n$, $n = 1,2,\ldots$. Consider then the sets defined by: $G_p = \{t \ ; \ \| V(t)x_o \| \leq p\}$, $t \in [-T,T]$. We see that $G_1 \subset G_2 \subset \ldots$ and $\bigcup_{p=1}^{\infty} G_p = [-T,T]$; also, meas $G_p \to 2T$ if $p \to \infty$.

Let us fix p_o sufficiently large in order that meas $G_{p_o} > \alpha > 0$. Then

$$\| V(t)x_o \| \leq p_o , \quad \forall t \in G_{p_o} = F \ \text{(by definition)}.$$

Let be $E_n = \{t_n - \tau, \tau \in F\}$. One deduces that

$$n \leq \| V(t_n)x_o \| = \| V(t_n-\tau)V(\tau)x_o \| \leq \| V(\tau)x_o \| \, \| V(t_n-\tau) \| \leq$$

$$p_o \| V(t_n-\tau) \| \quad .$$

One has therefore: $\| V(\sigma) \| \geq \dfrac{n}{p_o}$ if $\sigma \in E_n$. Let be $\bar{E} = \limsup E_n$; if $\sigma \in \bar{E}$, then σ belongs to an infinity of E_n, hence $\| V(\sigma) \| = \infty$. Also, meas $(\bigcup_1^{\infty} E_n) \leq 4T$ in view of the following: $\xi \in E_n \Rightarrow |\xi| \leq |t_n| + |\tau| \leq T + T = 2T$; hence $E_n \subset [-2T,2T]$ and $\bigcup_1^{\infty} E_n$ too.

In that case $\bar{E}$ is a non-empty set: in fact: $\bar{E} = \bigcap_1^{\infty} F_n$, where

$$F_n = \bigcup_{q=n}^{\infty} E_q \ ; \ F_1 \supset F_2 \supset \ldots ; \ \text{meas } F_1 \leq 4T \ ; \ \text{meas } \bar{E} = \lim_{n\to\infty} \text{meas } F_n \ ;$$

31

$F_n \supseteq E_n$, meas $F_n \geq$ meas E_n = meas $F > \alpha > 0$; hence: meas $E \geq \alpha > 0$, E cannot be empty. The existence of points σ where $\|V(\sigma)\|$ is arbitrarily large is a contradiction which proves the Theorem 6.1.

<u>Remark 1</u>: One-parameter groups of operators in Banach spaces were first studied by Gelfand in [12].

An important consequence of the above Theorem is given in

<u>Corollary 1</u>: *The operator group $V(t)$ is strongly continuous $\forall t \in R$, that is $V(t)x$ is an E-continuous function, $\forall x \in E$.*

A subsequent consequence is found in

<u>Corollary 2</u>: *Any solution of (6.1) - (6.2) (for an well-posed problem) has a strongly continuous derivative , $\forall t \in R$.*

In fact, if $u(t) = U(t)u(0)$, $u'(t) = \lim\limits_{h \to 0} U(t)\frac{1}{h}[u(h)-u(0)] =$

$\lim\limits_{h \to 0} V(t)\frac{1}{h}[u(h) - u(0)] = V(t)u'(0) = V(t)(Au(0))$, and apply Corollary 1.

One may deduce also (as in Ch. I, §4), the commutativity relation:
$AV(t)x = V(t)Ax$, $\forall x \in D(A)$, $\forall t \in R$.

Let us define the infinitesimal generator for groups of operators. Let be $\tilde{D} = \{x \in E$, $\lim\limits_{\eta \to 0} \frac{1}{\eta}(V(\eta)x - x)$ exists in $E\}$. Let be $\tilde{A}$, $\tilde{D} \to E$, given by $\tilde{A}x = \lim\limits_{\eta \to 0} \frac{1}{\eta}(V(\eta)x - x)$. We call this operator $\tilde{A}$ the infinitesimal generator of the operator group $V(t)$.

We have the following

<u>Proposition 1</u>: *If $x \in \tilde{D}$, the function $V(t)x$ is continuously differentiable, $\forall t \in R$.*

In fact, if $h \to 0$, we have: $\frac{1}{h}[V(t+h) - V(t)]x = V(t)\frac{1}{h}[V(h)-I]x \to V(t)\tilde{A}x,$

32

considered that $V(t) \in L(E,E)$. Accordingly, the derivative $\frac{d}{dt} V(t)x$ is E-continuous, $\forall t \in R$. This proves Proposition 1.

<u>Remark 2</u>: If, again, $x \in \tilde{D}$, then $\frac{V(h) - I}{h} V(t)x = V(t) \frac{1}{h} [V(h) - I]x$ which $\to V(t)\tilde{A}x$ as $h \to 0$. Therefore, $V(t)x \in \tilde{D}$, and $\tilde{A}V(t)x = V(t)\tilde{A}x$, $x \in \tilde{D}$, a commutativity property which has been also proved for semi-groups (Ch. I, §4, Remark 2).

Besides, we have $A \subset \tilde{A}$ so that $\tilde{A}$ is an extension of A (see [33-I] - page 250, for the definition); in fact, if $x \in D(A)$, $V(t)x = U(t)x = u(t)$, a solution of $u' = Au$ such that $u(0) = x$. Hence $u'(0) = \lim\limits_{h \to 0} \frac{U(h)x-x}{h}$ exists and $= Au(0)$, so that $x \in D(\tilde{A})$ and $\tilde{A}x = Ax$.

<u>Remark 3</u>: From the above considerations (in particular from Theorem 6.1) we infer that the Cauchy problem (6.1) - (6.2) is uniformly well-posed on the whole real line that is

If $u_n'(t) = Au_n(t)$, $t \in R$ and if $u_n(0) \to \theta$ as $n \to \infty$, then $u_n(t) \to \theta$ as $n \to \infty$, uniformly on any compact interval $[a,b] \subset R$.

Quite important, even if really simple, is the following

<u>Proposition 2</u>: *Let be $x_o \in E$ and $t_o \in R$ such that $V(t)x_o$ is a differentiable function for $t = t_o$. Then, necessarily, x_o belongs to $D(\tilde{A})$.* (If $D(\tilde{A})$ is E , then $\tilde{A}$ is continuous on E by the closed graph theorem: hence, $\forall (x_n)_1^\infty \subset D(A)$ such that $x_n \to x \in E$, $(Ax_n)_1^\infty$ is a Cauchy sequence convergent to $\tilde{A}x$; this is not the case for "most" operators A so that "often" $D(\tilde{A})$ is $\neq E$; this is important in connection with Remark 1, Section 5.)

33

<u>Proof of Proposition 2:</u> We assume existence of the limit:

$$\lim_{h \to 0} \frac{1}{h} \, [V(t_o + h) - V(t_o)]x_o \ .$$

So results existence of:

$$\lim_{h \to 0} V(-t_o)\frac{1}{h} \, [V(t_o+h) - V(t_o)]x_o = \lim_{h \to 0} \frac{1}{h} \, [V(h) - I]x_o \quad \text{so that} \quad x_o \in \tilde{D} = $$

$$D(\tilde{A}) \ .$$

Quite illuminating is the next

<u>Theorem 6.2:</u> *If the Cauchy problem: $u'(t) = Au(t)$, $u(0) = u_o \in D(A)$, is well-posed on the real line, the Cauchy problem: $v'(t) = -Av(t)$, $v(0) = v_o \in D(A)$ is also well-posed on the real line.*

(i) <u>Existence:</u>   Let $\overset{\vee}{U}(t)$, a family of operators, $D(A) \to D(A)$, be defined by $\overset{\vee}{U}(t)x = U(-t)x$, $\forall x \in D(A)$.   Let us put: $v(t) = \overset{\vee}{U}(t)v_o$. We obtain

$$v'(t) = \lim_{h \to 0} \frac{1}{h} \, [\overset{\vee}{U}(t+h) - \overset{\vee}{U}(t)]v_o = \lim_{h \to 0} \frac{1}{h} \, [U(-t-h) - U(-t)]v_o =$$

$$- \lim_{h \to 0} U(-t) \, \frac{U(-h)-I}{-h} \, v_o = - \lim_{h \to 0} \frac{U(-h)-I}{-h} \, U(-t)v_o = - \lim_{h \to 0} \frac{U(-h)-I}{-h} \, v(t) =$$

$$- Av(t) , \quad v(0) = v_o \ .$$

(ii) <u>Uniqueness:</u> Let be $w(t)$ a solution to: $w' = -Aw$, $w(o) = \theta$.   Let be $u(t) = w(-t)$, $t \in R$.   It follows: $u'(t) = -w'(-t) = -(-Aw(-t)) = Au(t)$, and $u(0) = \theta$.   Hence, $u(t) = \theta \ \forall t \in R$, which implies $w(t) = \theta \ \forall t \in R.$

(iii) <u>Continuous dependence:</u>   We see that any solution is given by: $v(t) = U(-t)v_o$, so that we can apply the properties of $U(t)$ and of its

34

extension $V(t)$.

Closely related to the above discussion is the Cauchy problem on a half-line for a couple of differential equations

$$u'(t) = Au(t) \, , \quad 0 \le t < \infty \, , \quad v'(t) = -Av(t) \, , \quad 0 \le t < \infty \qquad (6.7)$$

where A and $-A$ have the same domain of definition, D, which is dense in E . We assume that the initial value problems: $u(0) = u_o \, \epsilon \, D$, $v(0) = v_o \epsilon D$ are both well-posed and let us consider the semi-groups $V^+(t)$, $V^-(s)$ associated according to Chapter I. We then prove following

Theorem 6.3: *The commutativity relation:* $V^+(t)V^-(s)x = V^-(s)V^+(t)x$ *is satisfied,* $\forall x \, \epsilon \, E$, $\forall s, t \, \epsilon \, [o, \infty)$.

It is obviously sufficient to consider $x \, \epsilon \, D$ and then replace $V^-(s)$, $V^+(t)$ by their restriction $U^-(s)$, $U^+(t)$, which are mappings of D into itself. Hence, we shall prove: $U^+(t)U^-(s)x = U^-(s)U^+(t)x \ \forall x \, \epsilon \, D$. We have: $\dfrac{d}{dt} U^+(t)(U^-(s)x) = A(U^+(t)U^-(s)x) \, , \ 0 \le t < \infty \, , \ \forall$ fixed $s \, \epsilon \, [o, \infty)$ (by (6.7) applied to $u(t) = U^+(t) \, (U^-(s)x)$).

On the other hand, we shall use $V^-(s) \, \epsilon \, L(E;E)$ in order to prove $\dfrac{d}{dt} V^-(s) \, (U^+(t)x) = V^-(s) \dfrac{d}{dt} (U^+(t)x) = V^-(s)AU^+(t)x$. We know also (Ch. I, §4, Remark 1), that $V^-(s) \, (-Ay) = -AV^-(s)y \ \forall y \, \epsilon \, D$, so that

$$\dfrac{d}{dt} \, V^-(s) \, (U^+(t)x) = AV^-(s) \, U^+(t)x = AU^-(s)U^+(t)x \ .$$

Hence, both functions: $U^+(t)U^-(s)x$, $U^-(s)U^+(t)x$, $x \, \epsilon \, D$, verify the same differential equation (the first one in (6.7) (with respect to the variable t) and they have same initial data (for $t = 0$) , which equals $U^-(s)x$. Using uniqueness of the Cauchy problem we end the proof.

A very interesting result is obtained in the case of *uniformly* well-posed Cauchy problems

<u>Theorem 6.4</u>: *Let us assume that the Cauchy problem for both equations in (6.7) is uniformly well-posed. Then one has the relation $V^+(t)V^-(t)x = V^-(t)V^+(t)x = x$, $\forall x \in E$ and consequently $V^+(t) = [V^-(t)]^{-1}$, $V^-(t) = [V^+(t)]^{-1}$, $\forall t \geq 0$.*

<u>Proof</u>: Let us consider the function, $[0,\infty) \to E$, $z(t) = V^+(t)V^-(t)x_o$, where $x_o \in D$. For $t > 0$ we consider the expression $\frac{1}{\delta}[z(t+\delta) - z(t)]$ $(\delta > 0)$. It equals: $\frac{1}{\delta}[V^+(t+\delta)V^-(t+\delta)x_o - V^+(t)V^-(t)x_o] =$

$\frac{1}{\delta}[V^+(t+\delta) - V^+(t)]V^-(t)x_o + V^+(t+\delta)\frac{1}{\delta}[V^-(t+\delta) - V^-(t)]x_o$.

It is easy to see that: $\lim_{\delta \to 0} \frac{1}{\delta}[z(t+\delta) - z(t)] = AV^+(t)V^-(t)x_o + V^+(t)(-AV^-(t)x_o) = \theta$ (A commutes with V^+ on D); hence $z'(t) = \theta$ $\forall t > 0$, which implies $z(t) = z(\epsilon)$ $\forall t > 0$ and $0 < \epsilon < t$, that is $V^+(t)V^-(t)x_o = V^+(\epsilon)V^-(\epsilon)x_o$; if $\epsilon \to 0$ we see that $V^+(\epsilon)V^-(\epsilon)x_o \to x_o$ ($\forall x_o \in D$) so that $V^+(t)V^-(t)x_o = x_o$ $\forall x_o \in D$, hence $\forall x \in E$.

An important implication of Theorem 6.4: if the Cauchy problem for the couple of equations (6.7) is uniformly well-posed on $[0,\infty)$ then it is in fact uniformly well-posed on the whole real line for each equation in (6.7). In order to establish this fact, we shall first give the

<u>Lemma 6.1</u>: *The one-parameter families of operators: $W(t) = V^+(t)$ for $t \geq 0$, $W(t) = V^-(-t)$, for $t \leq 0$ and $Z(t) = V^-(t)$ for $t \geq 0$, $Z(t) = V^+(-t)$, for $t \leq 0$ as functions, $t \in R \to L(E,E)$, are strongly-continuous groups.*

It suffices to show, for example, that for $t > 0$ and $\tau < 0$ the group property: $W(t+\tau) = W(t)W(\tau)$, is satisfied. If $t + \tau > 0$ we have

36

$t = t + \tau - \tau$, hence $W(t) = V^+(t) = V^+(t + \tau - \tau) = V^+(t + \tau)V^+(-\tau)$ and

multiplying at right by $W(\tau)$, we obtain: $W(t)W(\tau) = V^+(t+\tau)V^+(-\tau)V^-(-\tau) = V^+(t+\tau)$ (after use of Theorem 6.4) $= W(t+\tau)$. If $t + \tau < 0$ we have:

$\tau = (\tau+t) - t$ so that $W(\tau) = V^-(-\tau) = V^-(-(\tau+t) + t) = V^-(t)V^-(-\tau-t)$;

multiply by $W(t)$ at left and get: $W(t)W(\tau) = V^+(t)V^-(t)V^-(-\tau-t) = V^-(-\tau-t) = W(\tau+t)$.

The strong continuity of $W(t)$ and $Z(t)$ for $t = 0$ is an obvious Corollary of this same property (at right) for $V^+(t)$ and $V^-(t)$.

We can end now by giving the

<u>Theorem 6.5</u>: *The Cauchy problem for each equation in (6.7) is uniformly well-posed on the real line.*

1. <u>Uniqueness</u>: Let us assume: $u'(t) = Au(t)$, $t \in R$, $u(0) = \theta$. By assumption we infer that $u(t) = \theta$ for $t \geq 0$. Let be $w(t) = u(-t)$, $t \geq 0$. Then $w'(t) = -u'(-t) = -Au(-t) = -Aw(t)$, $t \geq 0$, which implies also by assumption, that $w(t) = \theta$ for $t \geq 0$, that is $u(t) = \theta \ \forall t \leq 0$.

In a similar way we treat the second equation.

2. <u>Existence</u>: For the first equation. Let be $u_o \in D$ and put $u(t) = W(t)u_o$, so that for $t \geq 0$, $u(t) = U^+(t)u_o$, while for $t \leq 0$, $u(t) = U^-(-t)u_o$. Computing the derivative of $u(t)$ we find: $u'(t) = AU^+(t)u_o$, $t > 0$ and also $u'(t) = AU^-(-t)u_o$ for $t \leq 0$; this second equality follows from the fact that $\frac{d}{d\xi}U^-(\xi)x = -AU^-(\xi)x$, $\xi \geq 0$, $x \in D$, after changing variable: $t = -s$, $u(-s) = U^-(s)u_o$, $s \geq 0$; $v(s) = u(-s) \Rightarrow v'(s) = -u'(-s) = -AU^-(s)u_o$, $\Rightarrow u'(-s) = AU^-(s)u_o$, $u'(t) = AU^-(-t)u_o$, $t \leq 0$. For the second equation we put, given $v_o \in D$, $v(t) = Z(t)v_o$, and verify the relation $v' = -Av \ \forall$ real t in a similar

way.

The continuous dependence on the initial data is an obvious corollary of these formulas for the solution.   This ends the proof of Theorem 6.5.

§7.   <u>THE FUNDAMENTAL EXISTENCE THEOREMS</u>

In this section we shall explain some fundamental existence theorems for the uniformly well-posed Cauchy problem, $u'(t) = Au(t)$, $u(o) = u_o \in D(A)$ on the positive half-line $[o,\infty)$ and on the whole real line.

We shall first deduce necessary conditions which are strong enough in order to be also sufficient, i.e. to produce existence theorems.   Let us start with the Cauchy problem on the half-line $[o,\infty)$.   We assume that $D(A)$ is dense in E and that the Cauchy problem on $D(A)$ is uniformly well-posed.   We have seen that we can associate a C_o-semi-group $V(t)$ to this problem so that any solution $u(t)$ is given by $u(t) = V(t)u_o$.   If $\tilde{A}$ is the infinitesimal generator of $V(t)$ then (Theorem 4.1), $\tilde{A}$ is the closure of A .   Actually, the necessary conditions that we are able to produce bear on $\tilde{A}$ rather than on A .   In order to avoid new difficulties we shall assume from now on that: A is a closed operator, so that actually A will coincide with the infinitesimal generator of $V(t)$, $\tilde{A}$.   We shall therefore obtain some necessary conditions on this generator (precisely, on its resolvent operator $(\lambda I - \tilde{A})^{-1}$) which are a consequence of the formula

$$(\lambda I - \tilde{A})^{-1}x = \int_o^\infty e^{-\lambda t} V(t)x\,dt \; , \quad \forall x \in E \; ,$$

and which will be shown to be sufficient for the required existence result.   We have seen (Theorem 3.1), that the above integral is convergent for all complex λ with real part Re λ greater than ω .

We shall now prove

Theorem 7.1:   *Let be A a linear closed operator with dense domain $D(A) \subset E$*

and let us assume the Cauchy problem: $u'(t) = Au(t)$, $0 \leq t < \infty$,

$u(o) = u_o \ \varepsilon \ D(A)$ to be uniformly well-posed.   Then, there exist two numbers

$M > 0$ and ω real, such that, for any real $\lambda > \omega$, the resolvent operator

$(\lambda-A)^{-1} = R(\lambda;A)$ exists and belongs to $L(E,E)$, and furthermore, the

sequence of estimates: $\| [R(\lambda,A)]^n \| \, L(E,E) \leq \dfrac{M}{(\lambda-\omega)^n}$, $n = 1,2,\ldots$ is

satisfied.

Proof: Let be $V(t)$ the associated C_o-semi-group, so that A is its

infinitesimal generator, and the formula: $(\lambda-A)^{-1}x = \int_o^\infty e^{-\lambda t} \, V(t)x dt$ is

true, for $\lambda > \omega$ where ω appears in the estimate: $\| V(t) \| \leq Me^{\omega t}$, which

was proved in Theorem 2.1.

We shall derive a formula for the iterates of $R(\lambda;A)$, precisely the

following: $[R(\lambda;A)]^n x = \dfrac{1}{(n-1)!} \int_o^\infty e^{-\lambda t} \, t^{n-1} \, V(t)x dt$, $\forall \lambda > \omega$.

This will be a consequence of the formula for the n-th derivative of

$R(\lambda;A)$ with respect to $\lambda : \dfrac{d^n}{d\lambda^n} R(\lambda;A) = (-1)^n \cdot n! [R(\lambda;A)]^{n+1}$.   We can

prove this formula by induction with respect to n .   For $n = 1$, we

consider the ratio $\dfrac{1}{\lambda-\mu}[R(\lambda;A) - R(\mu;A)]$.   It is easy to see ([37] - page 211)

that this equals $-R(\lambda,A)R(\mu;A)$.

Regularity properties of $R(\lambda;A)$ with respect to λ can also be obtained

from the series: $R(\lambda;A) = R(\lambda_o;A) \{I + \sum\limits_{n=1}^\infty (\lambda_o-\lambda)^n [R(\lambda_o;A)]^n\}$, which is

convergent for $|\lambda - \lambda_o| < (\| R(\lambda_o;A) \|)^{-1}$, so that we deduce

$\lim\limits_{\mu \to \lambda} \dfrac{1}{\mu-\lambda} [R(\mu;A) - R(\lambda;A)] = -[R(\lambda;A)]^2$.

Now, in order to effectuate the passage from n to $n+1$, we consider

$$[R(\lambda;A)]^{(n)} - [R(\mu;A)]^{(n)} = (-1)^n \cdot n! \{[R(\lambda;A)]^{n+1} - [R(\mu;A)]^{n+1}\}$$

$$= (-1)^n n! [R(\lambda;A) - R(\mu;A)]\{ [R(\lambda;A)]^n + [R(\lambda;A)]^{n-1} R(\mu;A) + \ldots$$

$$+ [R(\mu;A)]^n\} \ .$$

Dividing by $\lambda - \mu$, and letting again μ to tend to λ, we obtain

$$[R(\lambda;A)]^{(n+1)} = (-1)^n . n! (-1) [R(\lambda;A)]^2 (n+1) [R(\lambda;A)]^n$$

$$= (-1)^{n+1} (n+1)! [R(\lambda;A)]^{n+2} \ .$$

Afterwards, we see without difficulty that the n-th derivative with respect to λ of the integral $\int_0^\infty e^{-\lambda t} V(t)x\,dt$, equals: $\int_0^\infty (-t)^n e^{-\lambda t} V(t)x\,dt$ so that

$$[R(\lambda;A)]^n x = \frac{1}{(n-1)!} (-1)^{n-1} [R(\lambda;A)]^{(n-1)} x$$

$$= \frac{1}{(n-1)!} (-1)^{n-1} \int_0^\infty (-t)^{n-1} e^{-\lambda t} V(t)x\,dt$$

$$= \frac{1}{(n-1)!} \int_0^\infty t^{n-1} e^{-\lambda t} V(t)x\,dt \ ;$$

if we now estimate $(\lambda > \omega)$, we get

$$\| [R(\lambda;A)]^n x \| \leq \frac{1}{(n-1)!} \int_0^\infty t^{n-1} e^{-\lambda t} Me^{\omega t} \| x \| \, dt = \frac{M}{(\lambda-\omega)^n} \| x \|$$

as easily seen.

This ends the proof of Theorem 7.1.

The converse fundamental result (Theorem of Hille-Yosida-Phillips) is here stated in the form of

<u>Theorem 7.2</u>: *Let be* A *a linear closed operator with dense domain* $D(A)$ *in the Banach space* E . *Let us assume existence of two numbers* $M > 0$ *and* ω *real, such that,* $\forall \lambda > \omega$, $(\lambda I - A)^{-1} \in L(E,E)$ *and* $\| [R(\lambda;A)]^n \|_{L(E,E)}$

$$\leq \frac{M}{(\lambda-\omega)^n} \quad , \quad n = 1,2,\ldots \quad .$$ *Then the Cauchy problem:* $u'(t) = Au(t)$,

$0 \leq t < \infty$, $u(0) = u_o \in D(A)$ *is uniformly well-posed.*

<u>Proof</u>: In view of Theorem 4.2, it will be sufficient to show that A is the infinitesimal generator of a semi-group $T(t)$ of class C_o . This is done here in a few steps.

Let us remember that, if B is a bounded linear operator in E , the exponential function: $\exp(tB)$ is well-defined as infinite series:

$\exp(tB) = \sum\limits_{n=0}^{\infty} \frac{1}{n!} \, t^n B^n$ which is norm-convergent (in $L(E,E)$) ; the estimate

$\| \exp(tB) \| \leq \exp(t\|B\|)$ is obviously verified and the differential equation

$\frac{d}{dt} [\exp(tB)]x = B \exp(tB)x$, $\forall t \in R$, $\forall x \in E$ is also true.

We shall now consider a family of bounded operators in E , B_λ , which tends to A in a definite sense when $\lambda \to \infty$; then we consider the exponential function: $\exp(tB_\lambda)$, and prove that it has also a definite limit: $T(t)$, when $\lambda \to \infty$ which is in fact the (unique) semi-group of class C_o whose infinitesimal generator is A . Precisely, let us define, for λ real $> \omega$, $B_\lambda = -\lambda[I-\lambda(\lambda I-A)^{-1}]$. Then, we see readily (because if $KL = LK$, $\Rightarrow \exp(tL)\exp(tK) = \exp(t(L+K))$) that: $\exp[tB_\lambda] = e^{-\lambda t}\exp[t\lambda^2(\lambda I-A)^{-1}] =$
$e^{-\lambda t} \sum\limits_{n=0}^{\infty} \frac{(t\lambda^2)^n}{n!} \, [(\lambda-A)^{-n}]$. Accordingly we have also the estimate, (in $L(E,E)$)

$$\| \exp[tB_\lambda] \| \leq e^{-\lambda t} \sum_{n=0}^{\infty} (\frac{t\lambda^2}{\lambda-\omega})^n \cdot \frac{M}{n!} = Me^{\frac{\lambda t\omega}{\lambda-\omega}} \quad .$$

Let us take an arbitrary number $\omega_1 > \omega$; as $\lim\limits_{\lambda\to\infty} \frac{\lambda\omega}{\lambda-\omega} = \omega$, we see that

$\frac{\lambda\omega}{\lambda-\omega} < \omega_1$ for $\lambda \geq \lambda_o$, hence, for $t \geq 0$ it is $\frac{\lambda t\omega}{\lambda-\omega} \leq t\omega_1$

$$\| \exp[tB_\lambda] \| \leq Me^{t\omega_1} \, , \quad \lambda \geq \lambda_o \, , \quad t \geq 0 \quad .$$

On the other hand, let us indicate that $\lim\limits_{\lambda\to\infty} B_\lambda x = Ax \ \forall x \in D(A)$. In fact, we see firstly that: $\lambda(\lambda I - A)^{-1}(\lambda I - A)x = \lambda x$, $\forall x \in D(A)$. Hence $\lambda R(\lambda:A)Ax = \lambda^2 R(\lambda;A)x - \lambda x = B_\lambda x$, $x \in D(A)$, so that it will suffice to show that: $\lim\limits_{\lambda\to\infty} \lambda R(\lambda:A)y = y \ \forall y \in E$. Actually, this is true for $y \in D(A)$ because in that case: $\lambda R(\lambda;A)y - y = \lambda R(\lambda,A)y - R(\lambda,A)(\lambda-A)y = R(\lambda;A)(Ay)$, so that: $\| \lambda R(\lambda;A)y - y \| = \| R(\lambda;A)(Ay) \| \leq \frac{M}{\lambda-\omega} \| Ay \| \to 0$ as $\lambda \to \infty$.

This is also true for any $y \in E$ because of density of $D(A)$ in E combined with the uniform estimate: $\| \lambda R(\lambda;A) \| \leq \frac{M\lambda}{\lambda-\omega} \leq M+1$ for $\lambda \geq \lambda_1$. Let us put now $T_\lambda(t) = \exp[tB_\lambda]$ and let us prove existence of the limit $\lim\limits_{\lambda\to\infty} T_\lambda(t)x$, $\forall x \in D(A)$, uniformly for t in any compact subinterval of $[o,\infty)$. Consider the difference $T_\lambda(t)x - T_\mu(t)x$ which is expressed (because of commutativity of operators $T_\lambda(t)$, $T_\mu(t)$, B_λ , B_μ : corollary of relation $(\lambda-A)^{-1}(\mu-A)^{-1} = (\mu-A)^{-1}(\lambda-A)^{-1}$), under the equality

$$T_\lambda(t)x - T_\mu(t)x = \int_0^t \frac{d}{ds}[T_\mu(t-s)T_\lambda(s)]x\,ds = \int_0^t T_\mu(t-s)T_\lambda(s)(B_\lambda - B_\mu)x\,ds \ .$$ Hence, the estimate

$$\| T_\lambda(t)x - T_\mu(t)x \| \leq \int_0^t Me^{(t-s)\omega_1}Me^{s\omega_1} \| (B_\lambda - B_\mu)x \| \, ds \leq M^2 te^{t\omega_1} \| (B_\lambda - B_\mu)x \|,$$

for $x \in D(A)$, $\lambda,\mu \geq \lambda_o$ which implies $\| T_\lambda(t)x - T_\mu(t)x \| \leq C_{\bar{t}} \| (B_\lambda - B_\mu)x \|$ when $0 \leq t \leq \bar{t}$. It follows that: $\lim\limits_{\lambda\to\infty} T_\lambda(t)x$ exists $\forall x \in D(A)$; let be $T(t)$, $D(A) \to E$ be defined by this limit: $T(t)x = \lim\limits_{\lambda\to\infty} T_\lambda(t)x$, $\forall x \in D(A)$. We obtain then $\| T(t)x - T_\mu(t)x \| \leq C_{\bar{t}} \| (A-B_\mu)x \|$, $x \in D(A)$, $0 \leq t \leq \bar{t}$, so that $T(t)x$ is uniform limit on $[o,\bar{t}]$ of $T_\mu(t)x$, and is accordingly, a continuous function of $t \geq 0$, $\forall x \in D(A)$. We extend $T(t)$ to the whole of E , as a linear continuous operator, using density of $D(A)$ in E and the uniform bound $\| T_\lambda(t) \| \leq Me^{t\omega_1}$, $\lambda \geq \lambda_o$, through:

$T(t)x = \lim_{\mu \to \infty} T_\mu(t)x$, $\forall x \in E$. We see that, from $\| T_\mu(t)x \| \leq Me^{t\omega_1} \| x \|$,

$\mu \geq \lambda_o$, it follows $\| T(t)x \| \leq Me^{t\omega_1} \| x \|$. If $x \in E$ and $(x_n)_1^\infty \subset D(A)$

such that $x_n \to x$ then $\| T(t)x - T(t)x_n \| \leq Me^{t\omega_1} \| x - x_n \| \to 0$ uniformly on

$t \in [o, \bar{t}]$, so that $T(t)x$ is continuous $\forall x \in E$ as well.

At this stage we can check that $T(t)$ is a (strongly continuous) operator

semi-group, having A as infinitesimal generator. We see that

$T(0)x = \lim_{\lambda \to \infty}[e^{tB_\lambda}x]_{t=o} = x \; \forall x \in E$, so that $T(0) = I$; next,

$T(t_1 + t_2)x = \lim_{\lambda \to \infty} \exp[(t_1 + t_2)B_\lambda]x = \lim_{\lambda \to \infty} \exp(t_1 B_\lambda)\exp(t_2 B_\lambda)x = T(t_1)T(t_2)x$,

this because $[T_\lambda(t_1)T_\lambda(t_2) - T(t_1)T(t_2)]x = [T_\lambda(t_1) - T(t_1)]T(t_2)x$

$+ T_\lambda(t_1)[T_\lambda(t_2)x - T(t_2)x]$ and from the estimate

$$\| T_\lambda(t_1)T_\lambda(t_2)x - T(t_1)T(t_2)x \| = \| [T_\lambda(t_1)-T(t_1)]T(t_2)x \| + \| T_\lambda(t_1) \| \cdot$$

$\| T_\lambda(t_2)x - T(t_2)x \|$ and the uniform bound: $\| T_\lambda(t_1) \| \leq Me^{t_1\omega_1}$, $\lambda \geq \lambda_o$.

Let us find now the infinitesimal generator of $T(t)$. Remark the relation

$T(t)x - x = \int_o^t T(s)Axds$, $\forall x \in D(A)$. In fact, we have obviously

$T_\lambda(t)x - x = \int_o^t T_\lambda(s)B_\lambda xds$ $(\frac{d}{ds} T_\lambda(s)x = T_\lambda(s)B_\lambda x)$, and when $\lambda \to \infty$ we see

that $\int_o^t T_\lambda(s)B_\lambda xds \to \int_o^t T(s)Axds$ for $x \in D(A)$; in fact

$$(T_\lambda(s)B_\lambda x - T(s)Ax) = T_\lambda(s)(B_\lambda x - Ax) + [T_\lambda(s) - T(s)]Ax$$

which tends to θ as $\lambda \to \infty$, uniformly for $0 \leq s \leq t$ because

$\| T_\lambda(s) \| \leq M(\sup(e^{\omega_1 t}, 1))$, $0 \leq s \leq t$, $\lambda \geq \lambda_o$, and of the uniform

convergence of $T_\lambda(s)y$ to $T(s)y$, on $0 \leq s \leq t$, $\forall y \in E$. (This

uniform convergence was only seen previously for $y \in D(A)$; in the general

case we use an estimate for: $T_\lambda(s)y - T(s)y = T_\lambda(s)[y-y_\epsilon] + T_\lambda(s)y_\epsilon -$

$T(s)y_\epsilon + T(s)y_\epsilon - T(s)y$, where $y_\epsilon \in D(A)$ and $\| y - y_\epsilon \| < \epsilon$, in the well-known way).   Hence, we can write: $\frac{1}{t}(T(t)x-x) = \frac{1}{t}\int_o^t T(s)Axds$, $x \in D(A)$, so

that, if we denote by B for one moment the infinitesimal generator of $T(t)$ we see that x is in $D(B)$ and $Bx = Ax$, hence, $B \supseteq A$.

Remark now that there exists a real λ such that $(\lambda-A)^{-1} \in L(E,E)$ and also $(\lambda-B)^{-1} \in L(E,E)$ (because of the assumption in Theorem 7.2 and from Theorem 7.1 and Theorem 4.2).   We shall prove that $D(B) = D(A)$, i.e. that $D(B) \subseteq D(A)$.   Let $x \in D(B)$, and $y = \lambda x - Bx$.   Then, $\exists x_1 \in D(A)$, such that $(\lambda-A)x_1 = y$.   Hence, $(\lambda-B)x_1 = (\lambda-B)x = y$ and $x = x_1 \in D(A)$. Therefore $A = B$, and Theorem 7.2 is proved.

<u>Remark</u>: For the above constructed semi-group $T(t)$ we obtained an estimate $\| T(t) \| \leq Me^{t\omega_1}$, $\forall t \geq 0$, where ω_1 is any number $> \omega$ and M is independent of it; we can deduce therefore the precise inequality: $\| T(t) \| \leq Me^{t\omega}$, with the same constants M and ω which appear in the assumptions of the theorem.

In the remaining part of this section we shall see how the above obtained existence results for uniformly well-posed Cauchy problem on a half-line can be applied in order to obtain also existence results for uniformly well-posed Cauchy problem on the whole real line (see §6 above).   We shall state and prove the following

<u>Theorem 7.3</u>: *Let be A a closed linear operator with dense domain $D(A)$ in the Banach space E .   A necessary and sufficient condition in order that the Cauchy problem, $u'(t) = Au(t)$, $-\infty < t < \infty$, $u(0) = u_o \in D(A)$ be uniformly well-posed, is the existence of two positive numbers $M > 0$, $\omega \geq 0$ such that for λ real, $|\lambda| > \omega$, $(\lambda I-A)^{-1} \in L(E,E)$ and $\| (\lambda I-A)^{-n} \| \leq \frac{M}{(|\lambda|-\omega)^n}$, $n = 1,2,\ldots$*

Proof of necessity: We first use section 6, in order to remark that the Cauchy problems for $u' = Au$ and $u' = -Au$ are both uniformly well-posed on $[o,\infty)$ (Theorem 6.2). Now apply Theorem 7.1. First find a real ω_1 such that $\lambda > \omega_1 \Rightarrow (\lambda-A)^{-1} \in L(E,E)$. Next find another real ω' such that $\lambda > \omega' \Rightarrow (\lambda+A)^{-1} \in L(E,E)$. Hence $\mu < -\omega' \Rightarrow -\mu > \omega' \Rightarrow (-\mu+A)^{-1} \in L(E,E)$. But, also $-(-\mu+A)^{-1} = R(\mu;A)$ because $-(-\mu+A)^{-1} \cdot (\mu-A) = (-\mu+A)^{-1}(-\mu+A) = I_{D(A)}$ and $(\mu-A) \times (-)(-\mu+A)^{-1} = (-\mu+A)(-\mu+A)^{-1} = I_E$. Hence, $(\mu-A)^{-1} = R(\mu;A) \in L(E,E)$ also for $\mu < -\omega'$; we can take therefore $\omega = \sup(|\omega_1|, |\omega'|)$. In fact, if $|\lambda| > \omega$, then, if $\lambda > 0$,

$$|\lambda| = \lambda > \omega \geq |\omega_1| \geq \omega_1, \quad \text{and if } \lambda < 0, \quad |\lambda| = -\lambda > |\omega'| \geq \omega' \Rightarrow \lambda < -\omega',$$

hence in both cases $(\lambda-A)^{-1} \in L(E,E)$; in what concerns the sequence of estimates: the above choosen ω_1 and ω' permit also, respectively, the estimates

$$\| [(\lambda-A)^{-1}]^n \| \leq \frac{M}{(\lambda-\omega_1)^n}, \quad \lambda > \omega_1, \; n = 1,2,\ldots$$

and

$$\| [(\lambda+A)^{-1}]^n \| \leq \frac{M}{(\lambda-\omega')^n}, \quad \lambda > \omega', \; n = 1,2,\ldots$$

Hence, again: if $\lambda > 0$, $|\lambda| > \omega \Rightarrow \lambda > \omega_1$ and we can write

$$\| [(\lambda-A)^{-1}]^n \| \leq \frac{M}{(|\lambda|-\omega_1)^n} \leq \frac{M}{(|\lambda|-\omega)^n}$$

as $\omega_1 \leq \omega$, while for $\lambda < 0$, $|\lambda| > \omega$,

$$\|[(\lambda-A)^{-1}]^n\| = \|(-1)^n[-\lambda+A)^{-1}]^n\| = \| [(-\lambda+A)^{-1}]^n \| \leq \frac{M}{(-\lambda-\omega')^n}$$

because $-\lambda > \omega'$, that is

$$\left\| \left[(\lambda - A)^{-1} \right]^n \right\| \; \leq \; \frac{M}{(|\lambda| - \omega')^n} \; \leq \; \frac{M}{(|\lambda| - \omega)^n}$$

as $\omega' \leq \omega$, $n = 1, 2, \ldots$

<u>Proof of sufficiency</u>: If λ is a positive number $> \omega$, then $\left\| (\lambda - A)^{-n} \right\| \leq \frac{M}{(\lambda - \omega)^n}$, $n = 1, 2, \ldots$: hence, by Theorem 7.2, we infer that the Cauchy problem on $[0, \infty)$ for $u' = Au$, $u(0) \in D(A)$ is uniformly well-posed. On the other hand we deduce that $(\lambda - A)^{-1} \in L(E, E)$ for $\lambda < -\omega$. But obviously $(\lambda - A)^{-1} = [-(-\lambda + A)]^{-1} = -(-\lambda + A)^{-1}$ so that, if $-\lambda = \nu$, we can say that $(\nu + A)^{-1}$ exists for $\nu > \omega$ and $(\nu + A)^{-1} = -(\lambda - A)^{-1}$. Consequently $(\nu + A)^{-n} = [-(\lambda - A)^{-1}]^n = (-1)^n (\lambda - A)^{-n}$, so that $\left\| (\nu + A)^{-n} \right\| = \left\| (\lambda - A)^{-n} \right\| \leq$

$$\frac{M}{(|\lambda| - \omega)^n} = \frac{M}{(-\lambda - \omega)^n} = \frac{M}{(\nu - \omega)^n} \; , \; n = 1, 2, \ldots \; , \quad \text{for } \nu > \omega \; .$$

This shows that Theorem 7.2 applies equally to the operator $-A$ so that the Cauchy problem on $[0, \infty)$ for $u' = -Au$, $u(o) \in D(A)$ is uniformly well-posed. The result is now a consequence of Theorem 6.5 and in fact the Cauchy problem for both equations $u' = \pm Au$, $u(0) \in D(A)$ is uniformly well-posed on the *whole* real line.

§8.   SOME ABSTRACT APPLICATIONS

In this section we shall give 3 applications of the previous section: the first two refer to existence results on $[0, \infty)$, the last one is a global existence result.   To start, we consider a Hilbert space H , where the norm comes from a scalar product: $\|x\| = (x, x)^{\frac{1}{2}}$, $\forall x \in H$ and then let us assume (as in the work of Liance [24]) that a linear closed operator A with dense domain $D(A) \subseteq H$ is given (so that its adjoint operator A^* (see [37] for definition) is also densely defined in H) ($D(A^*)$ is its domain)), and that a pair of inequalities

$$\mathrm{Re}(Ah,h) \leq \beta(h,h) \ , \quad \forall h \ \epsilon \ D(A), \ \mathrm{Re}(A^*k,k) \leq \beta(k,k) \ , \quad \forall k \ \epsilon \ D(A^*) \tag{8.1}$$

is satisfied for a certain (real) number β . In this case we prove the following (see Liance [24]).

<u>Theorem 8.1</u>: *The Cauchy problem for $u' - Au = o$, $u(o) \ \epsilon \ D(A)$ is uniformly well-posed on the positive half-axis.*

We shall check in fact the possibility of applying Theorem 7.2. Let us show, as a first step, that for any real number $\lambda > \beta$, the resolvent operator $(\lambda-A)^{-1}$ exists and $\epsilon \ L(H;H)$. Therefore, let us consider the equation $(\lambda-A)f = g$, where g is given in H . If we take scalar products we obtain: $\lambda(f,f) - (Af,f) = (g,f)$, and by taking also the complex-conjugate of this equality, we derive: $\bar{\lambda}(f,f) - \overline{(Af,f)} = (f,g)$. Summing both equalities we obtain $(2\mathrm{Re}\lambda)\| f\|^2 - 2\mathrm{Re}(Af,f) = (g,f) + (f,g) \leq 2\| f\| \| g\|$ (Schwarz inequality) hence, because of $\mathrm{Re}(Af,f) \leq \beta\| f\|^2$, we also obtain that $(\mathrm{Re}\lambda - \beta)\| f\|^2 \leq \| f\| \| g\|$, so that for λ real and $>\beta$ we get: $\| f\| \leq \frac{1}{\lambda-\beta} \| g\|$; accordingly, if $(\lambda-A)f = g = \theta$, it follows that $f = \theta$ too, therefore the existence of the inverse operator $(\lambda-A)^{-1}$ is proved for $\lambda > \beta$. We shall next prove that the range of $\lambda - A$; i.e., the set $R_{\lambda-A} = (\lambda-A)(D(A))$ is the entire space H . First we see that it is a closed linear subspace of H . As the linearity of $R_{\lambda-A}$ is obvious, let us assume that $g_n = (\lambda-A)f_n$, $f_n \ \epsilon \ D(A)$ and $g_n \to g_o \ \epsilon \ H$. Using an estimate just obtained, we see that $\| f_n-f_m\| \leq \frac{1}{\lambda-\beta} \| g_n-g_m\|$, $\lambda > \beta$, hence f_n is also convergent, say to f_o . Use now closedness of the operator $\lambda - A$; one obtains: $f_o \ \epsilon \ D(A), \ (\lambda-A)f_o = g_o$, so that

$g_o \in (\lambda-A)(D(A))$.   Let us assume now, reasoning by contradiction, that $R_{(\lambda-A)}$ is a strict subspace of H .   Then, one is able to find an element $h_o \in H$, $h_o \neq \theta$ and $(\lambda f - Af, h_o) = 0$, $\forall f \in D(A)$, that is $\lambda(f, h_o) = (Af, h_o)$.   At this point we can use the *definition* of the adjoint operator A^* , and obtain $h_o \in D(A^*)$, $A^* h_o = \lambda h_o$.   Let us compute then $Re(A^* h_o, h_o)$ which equals $\lambda \| h_o \|^2 > \beta \| h_o \|^2$; thus contradicting one hypothesis of the Theorem: therefore $(\lambda-A)(D(A)) = H$ so that $(\lambda-A)^{-1}$ is everywhere defined.   If $(\lambda-A)f = g$, then $\| f \| \leq \dfrac{1}{\lambda-\beta} \| g \|$ (for $\lambda > \beta$) and $f = (\lambda-A)^{-1} g$, hence $\| (\lambda-A)^{-1} \| \leq \dfrac{1}{\lambda-\beta}$ for $\lambda > \beta$.   This obviously implies $\| (\lambda-A)^{-n} \| \leq \dfrac{1}{(\lambda-\beta)^n}$, $n = 1, 2, \ldots$ and an application of Theorem 7.2 establishes the required result.

Next, we shall partially extend this line of reasoning from Hilbert to Banach spaces, and we shall prove in fact the following (simple)

<u>Theorem 8.2</u>: *Suppose X a Banach space; A a linear operator with dense domain $D(A) \subseteq X$, such that $\lambda \in (0, \infty) \Rightarrow (\lambda-A)^{-1} \in L(X, X)$.   Assume further that for any $x \in D(A)$ there exists a linear continuous functional F_x on X , such that: $\| F_x \|_{X^*} \leq \| x \|$, $F_x(x) = \| x \|^2$, $F_x(Ax) \leq 0$.   Then, the Cauchy problem: $u' - Au = \theta$, $u(0) \in D(A)$ on $[0, \infty)$ is uniformly well-posed.* (See references [38], [48], [49]); X^* is the dual space to X.

It is obviously sufficient to prove the estimate: $\| (\lambda-A)^{-1} \| \leq \dfrac{1}{\lambda}$, $\lambda > 0$.   Consider the equation $\lambda x - Ax = y$, y given in X , $\lambda > 0$.   To the solution $x = (\lambda-A)^{-1} y$ which exists by hypothesis, we associate, and apply, the functional F_x ; we obtain: $F_x(\lambda x) - F_x(Ax) = F_x(y)$, that is $\lambda F_x(x) - F_x(Ax) = F_x(y)$ and therefore: $\lambda \| x \|^2 \leq \| F_x \| \| y \| \leq \| x \| \| y \|$ hence $\| x \| \leq \dfrac{1}{\lambda} \| y \|$, $\forall y \in X$, which implies the required result.

As a last application we will solve the initial value problem for a second order equation $\dfrac{d^2 u}{dt^2} = Bu$, $u(o) = u_o$, $u'(o) = u_1$, in a Hilbert space,

transforming it into a first order equation $v' = Bv$ in a cartesian product

of Hilbert spaces and then applying Theorem 7.3 so that, actually, the

problem will be uniformly well-posed on the whole real line.   More precisely:

let be H a Hilbert space and let $B = -A^2$ where A and A^2 are self-

adjoint operators in H and $(A^2 h, h) \geq (h, h)$, $\forall h \in D(A^2)$.   Our Cauchy

problem will consist in finding a function $u(t)$, $R \rightarrow D(A^2)$, twice

continuously differentiable in H , verifying relations $u''(t) = -A^2 u(t)$,

$u(o) = u_o \in D(A^2)$, $u'(o) = u_1 \in D(A)$.   If we introduce the vector-function:

$v(t) = (u(t), u'(t))$, then we see that $v'(t) = A\,v(t)$ where A is the

matrix-operator $\begin{pmatrix} \theta & I \\ -A^2 & \theta \end{pmatrix}$, and $v(o) = \{u_o, u_1\} \in D(A^2) \times D(A)$.   We shall

formalize things, introducing a second Hilbert space, K , in the following

way: remark that $(A^2 h, h) = (Ah , Ah) \geq \| h \|^2$, $\forall h \in D(A^2)$.   Let us

introduce on the set $D(A)$ a new norm and scalar product by the formulas:

$(h, k)_K = (Ah, Ak)_H$, $h, k \in D(A)$, $\| h \|_K^2 = \| Ah \|_H^2$.   The set $D(A)$ is hence

a pre-Hilbert space.   It is also a complete space because: if $(h_n)_1^\infty$ is a

Cauchy sequence in $D(A)$, it will also be (in view of estimate:

$\| h_n - h_m \|_K \geq \| h_n - h_m \|_H$) a Cauchy sequence in H ;   hence $h_n \rightarrow h_o \in H$

while Ah_n is a Cauchy sequence in H .   But A is a closed operator so

that $h_o \in D(A)$ and $Ah_o = \lim\limits_{n \to \infty} Ah_n$;   we can say that $h_n \rightarrow h_o$ in $D(A)$,

$\| h_n - h_o \|_K \rightarrow 0$.   Hence, our second Hilbert space K is a subspace of H

with a stronger norm: $\| h \|_K = \| Ah \|_H \geq \| h \|_H$.   Let us consider now the

cartesian product $K \times H = \mathcal{H}$ where the scalar product is defined in the

natural way: for $\{u_1, v_1\}$, $\{u_2, v_2\}$ in $\mathcal{H}$, we put $(\{u_1, v_1\}, \{u_2, v_2\})_{\mathcal{H}} =$

$(u_1, u_2)_K + (v_1, v_2)_H$.   Hence $K \times H$ is again a Hilbert space.   Consider

the operator A which is defined on the set $D(A) = D(A^2) \times D(A)$ by the relation: $A\{u,v\} = \{v,-A^2u\}$, $\forall u \in D(A^2)$, $v \in D(A)$. We shall prove now that for any λ real, $\lambda \neq 0$, the inverse operator $(\lambda-A)^{-1}$ belongs to $L(H,H)$ and $\| (\lambda-A)^{-1} \| \leq \frac{1}{|\lambda|}$. Let us start by solving the equation $(\lambda-A)\{u,v\} = \{f,g\}$, for $\lambda \neq 0$ and $\{f,g\}$ given in H. This equation is in fact: $\{\lambda u, \lambda v\} - \{v,-A^2u\} = \{f,g\}$ or $\{\lambda u - v, \lambda v + A^2u\} = \{f,g\}$ which splits to the pair of equations: $\lambda u - v = f$, $\lambda v + A^2 u = g$. If u and v are solutions of this pair of equations, then necessarily $v = \lambda u - f$ and then $A^2 u + \lambda^2 u - \lambda f = g$ or $(A^2 + \lambda^2 I)u = \lambda f + g$. Here $\lambda f + g \in H$ so that the (unique) solution $u \in D(A^2)$ is given by: $u = (A^2 + \lambda^2 I)^{-1}(\lambda f + g)$ and consequently $v = (A^2 + \lambda^2 I)^{-1}(\lambda^2 f + \lambda g) - f$ $(v \in D(A)$ because $v = \lambda u - f$, $u \in D(A^2)$, $f \in D(A))$. (It is easy to check that u and v given by these formulas are actual solutions).

This shows that for $\lambda \neq 0$, the range of the operator $(\lambda-A)$, with domain $D(A) = D(A^2) \times D(A)$ is the whole space $H = K \times H$. In order to obtain the estimate $\| (\lambda-A)^{-1} \| \leq \frac{1}{|\lambda|}$ we shall consider, as in the preceding theorem, the expression $\mathrm{Re}(A\{u,v\}, \{u,v\})_H$. This equals in fact: $\mathrm{Re}(\{v,-A^2u\}, \{u,v\})_H = \mathrm{Re}[(v,u)_K - (A^2u,v)_H]$. As $v \in D(A)$, $(A^2u,v)_H = (Au,Av)_H = (u,v)_K$, and we see that $(v,u)_K - (u,v)_K = -2i\,\mathrm{Im}(u,v)_K$, hence $\mathrm{Re}(A\{u,v\}, \{u,v\})_H = 0$.

Reasoning in a similar way to Theorem 8.1 we obtain the required estimate. Then we can apply Theorem 7.3 and find that the Cauchy problem for the equation: $\frac{d}{dt}\{u,v\} = A\{u,v\}$, $\{u,v\}(0) = \{u_o,v_o\} \in D(A)$ is uniformly well-posed on the whole real line. This splits into the equations: $\frac{du}{dt} = v$, $\frac{dv}{dt} = -A^2 u$, $u(0) = u_o$, $v(0) = v_o$; it implies: $\frac{d^2u}{dt^2} = -A^2 u$, $u(0) = u_o$, $u'(0) = v_o$ giving a solution on the whole real line of the Cauchy problem here considered, which is uniformly well-posed in a natural sense. (See reference [40] for a special case).

3 Uniqueness of the Cauchy problem

<u>INTRODUCTION</u>

One considers solutions $u(t)$ of the first-order differential equation:
$\frac{du(t)}{dt} = Au(t)$, $0 \leq t \leq T$. We say that there is uniqueness of the initial-
value = Cauchy problem if the relation $u(0) = \theta$ implies: $u(t) = \theta$ for
$0 < t \leq T$. We have studied in the preceding two Chapters results concerning
existence, uniqueness and continuous dependence on the initial data. ("Well-
posed Cauchy problem"). Now we discuss briefly, following essentially the
fundamental work by Agmon-Nirenberg [2], the uniqueness of a Cauchy problem
which is not necessarily well-posed.

§1. <u>THE FIRST UNIQUENESS RESULT</u>

Let H be a Hilbert space with scalar product $(,)$ and associated norm
$\| \ \|$. Let us consider a linear closed operator A , which maps $D(A) \subseteq H$
into H , such that $D(A)$ be dense in H . One makes the following (see
reference [2] - Theorem 1.2, pag. 132).

<u>Hypothesis 1</u>: *There exists a sequence of vertical lines in the complex plane,*
$\{Re \ \lambda = \sigma_n\}_{n=1}^{\infty}$, $\sigma_n \to +\infty$ *as* $n \to \infty$ *with the following property: the*
resolvent operator $(\lambda - A)^{-1} = R(\lambda;A)$ *exists and belongs to* $L(H;H)$ *for*
$Re \ \lambda = \sigma_n$, $n = 1,2,\ldots$, *and the uniform bound:* $\| R(\lambda;A) \|_{L(H;H)} \leq M$, *for*
$Re \ \lambda = \sigma_n$, $n = 1,2,\ldots$, *is verified.*

<u>Remark</u>: It could have some interest to compare *Hypothesis 1* with Theorem 7.1
in Chapter II. For a uniformly well-posed Cauchy problem on $[0,\infty)$ with
linear closed-densely defined operator A , the resolvent exists for all

real $\lambda > \omega$ (and in fact it is easy to see that it exists in the whole half-plane Re $\lambda > \omega$) and verifies stronger than "uniformly bounded" estimates - so that $R(\lambda;A) \to \theta$ as $\lambda \to +\infty$. Hence, if the operator A considered in this Chapter has an uniformly bounded resolvent on a sequence of lines in Re $\lambda > \omega$, but *has no* continuous resolvent for a sequence of positive $\lambda_n \to +\infty$, we cannot expect well-posedness (i.e. existence and continuous dependence) even if we are able to prove uniqueness of the initial-value problem.

It would be nice to see some more concrete examples, and also, say, "existence without uniqueness" results but we are not ready to explain such facts in the present work.

We shall now state and prove the following

Theorem 1.1: *Let be A an operator verifying the Hypothesis 1.   Let also $u(t)$, $0 \leq t \leq T \to D(A)$ be a H-continuously differentiable function (right-derivative only for $t = 0$), verifying the equation $u'(t) = Au(t)$, $0 \leq t \leq T$ and the initial condition $u(0) = \theta$.   It follows that $u(t) = \theta$ for $0 \leq t \leq T$.*

Proof: Let us consider an auxiliary real-valued function $\xi(t) \in C^{\infty}(0,T)$, such that $\xi(t) = 1$ for $0 \leq t \leq \alpha < T$ and $\xi(t) = 0$ for $\alpha + \delta \leq t \leq T$, where $0 < \alpha < \alpha + \delta < T$.

Let us define then, for any real number σ , the H-valued function $v_{\sigma}(t)$, which equals: $e^{-\sigma t}\xi(t)u(t)$ for $t \in [0,T]$ and is null for $t < 0$ and for $t > T$.   Now, if $0 \leq t \leq \alpha$, we see that $v_{\sigma}(t) = e^{-\sigma t}u(t)$ and we see also that $v_{\sigma}(0) = \theta$.   Consequently we have: $\frac{1}{t} [v_{\sigma}(t) - v_{\sigma}(0)] =$ $\frac{1}{t} [e^{-\sigma t}u(t)] = e^{-\sigma t}[\frac{u(t) - u(0)}{t}]$, $0 \leq t \leq \delta$.   This has a limit when $t \downarrow 0$, namely $u'_+(0)$ which by hypothesis equals $Au(0) = \theta$.   Hence, $v_{\sigma}(t)$ has right-derivative in the origin = θ = left-derivative at $t = 0$.

52

follows that the derivative $v_\sigma'(0)$ exists and equals θ.

Hence, for any value of t, the function $v_\sigma(t)$ is easily seen to be continuously differentiable, and $v_\sigma'(t) = \theta$ for $t \leq 0$ and also for $t \geq T$. Furthermore, it is easy to check that: $v_\sigma'(t) = (A-\sigma I)v_\sigma(t) + e^{-\sigma t}\xi'(t)u(t)$ for all $t \in [0,T]$. Accordingly, we can write, after defining the function $f_\sigma(t)$ which vanishes for $t \notin [0,T]$ and equals $e^{-\sigma t}\xi'(t)u(t)$ for $0 \leq t \leq T$, the equality

$$v_\sigma'(t) + (\sigma I - A)v_\sigma(t) = f_\sigma(t) \tag{1.1}$$

on the whole real line: $-\infty < t < \infty$.

Let us consider now the Fourier transforms

$$\hat{v}_\sigma(\tau) = \frac{1}{\sqrt{2\pi}} \int_{-\infty}^{\infty} e^{-i\tau t} v_\sigma(t)dt \,, \quad \hat{f}_\sigma(\tau) = \frac{1}{\sqrt{2\pi}} \int_{-\infty}^{\infty} e^{-i\tau t} f_\sigma(t)dt \,;$$

their existence is obvious, for any real τ, because both v_σ and f_σ vanish outside $[0,T]$. We consider also the Fourier transform of the continuous derivative $v_\sigma'(t) : \widehat{v_\sigma'}(\tau) = \frac{1}{\sqrt{2\pi}} \int_{-\infty}^{\infty} e^{-i\tau t} v_\sigma'(t)dt$, and remark, by partial integration, the equality: $\widehat{v_\sigma'}(\tau) = i\tau\hat{v}_\sigma(\tau)$. In the equation (1.1) we shall multiply both sides by $\frac{1}{\sqrt{2\pi}} e^{-i\tau t}$, $\tau \in R$, and then we shall integrate on the real line. Using once more-say-the Theorem 1.3.5 [21], we can deduce from (1.1) the equality

$$[(\sigma + i\tau)I - A]\hat{v}_\sigma(\tau) = \hat{f}_\sigma(\tau) \,, \quad -\infty < \tau < \infty \,, \tag{1.2}$$

where $\hat{v}_\sigma(\tau) \in D(A)$ $\forall$ real value of τ. We make now essential use of the Hypothesis 1; in (1.2) we take $\sigma = \sigma_n$, $n = 1,2,\dots$, and derive, in an obvious manner, the representation formula

$$\hat{v}_{\sigma_n}(\tau) = [(\sigma_n + i\tau)I - A]^{-1}\,\hat{f}_{\sigma_n}(\tau)\,, \quad -\infty < \tau < \infty \tag{1.3}$$

and also the uniform estimate

$$\|\hat{v}_{\sigma_n}(\tau)\|_H \leq M\|\hat{f}_{\sigma_n}(\tau)\|_H\,, \quad -\infty < \tau < \infty\,, \quad n = 1,2,\dots\,. \tag{1.4}$$

We shall now use Parseval's equality for square-Bochner integrable H-valued functions

$$\int_{-\infty}^{\infty}\|h(t)\|_H^2\,dt = \int_{-\infty}^{\infty}\|\hat{h}(\tau)\|_H^2\,d\tau \tag{1.5}$$

where the Fourier transform $\hat{h}$ has a convenient[1] definition (reducing to the above written integral: $\hat{h}(\tau) = \dfrac{1}{\sqrt{2\pi}}\displaystyle\int_{-\infty}^{\infty} e^{-i\tau t}h(t)dt$ when $h(t)$ is Bochner integrable on the real line, as for example when it is continuous with compact support like $v_\sigma(t)$ and $f_\sigma(t)$).

We obtain then the relation

$$\int_{-\infty}^{\infty}\|v_{\sigma_n}(t)\|^2 dt = \int_{-\infty}^{\infty}\|\hat{v}_{\sigma_n}(\tau)\|^2 d\tau \leq M^2\int_{-\infty}^{\infty}\|\hat{f}_{\sigma_n}(\tau)\|^2 d\tau = M^2\int_{-\infty}^{\infty}\|f_{\sigma_n}(t)\|^2 dt$$

$$\tag{1.6}$$

and accordingly the inequalities

$$\int_{0}^{\alpha} e^{-2\sigma_n t}\|u(t)\|^2 dt \leq M^2\int_{\alpha}^{\alpha+\delta} |\xi'(t)|^2 e^{-2\sigma_n t}\|u(t)\|^2 dt$$

$$\leq M_1\int_{\alpha}^{\alpha+\delta} e^{-2\sigma_n t}\|u(t)\|^2 dt\,, \quad \forall n=1,2,\dots\,. \tag{1.7}$$

Let us take now a number $\beta > 0$ such that $\beta < \alpha$, and remark also that

---

[1]One may imitate definitions and proofs given for scalar-valued functions, they extend easily to Hilbert-space valued functions, or also, when H is separable, one may reduce to scalar case taking an orthonormal basis (see A. Friedman [11]).

$\sigma_n > 0$ for $n \geq n_o$; in that case, a corollary to (1.7) is the estimate

(after using: $\int_o^\beta e^{-2\sigma_n t} \|u(t)\|^2 dt \leq \int_o^\alpha e^{-2\sigma_n t} \|u(t)\|^2 dt$) .

$$e^{-2\sigma_n \beta} \int_o^\beta \|u(t)\|^2 dt \leq M_1 \, e^{-2\sigma_n \alpha} \int_\alpha^{\alpha+\delta} \|u(t)\|^2 dt \, , \quad n \geq n_o \qquad (1.8)$$

and also

$$\int_o^\beta \|u(t)\|^2 dt \leq M_1 \, e^{2\sigma_n(\beta-\alpha)} \int_\alpha^{\alpha+\delta} \|u(t)\|^2 dt \, , \quad n \geq n_o \qquad (1.9)$$

We see readily that (1.9) implies: $\int_o^\beta \|u(t)\|^2 dt = 0$; hence $u(t) = \theta$

for $0 \leq t \leq \beta$. But β can be arbitrarily close to α , so that

$u(t) = \theta$ on $[0,\alpha]$. Finally, the number α which was fixed in the above

reasoning, can be taken arbitrarily close to T , so that $u(t) \equiv \theta$ in $[0,T]$.

§2. UNIQUENESS VIA CONVEXITY

In this section we shall prove, under quite different assumptions, but

maintaining the hilbertian nature of the underlying space, a second uniqueness

result for the previously considered Cauchy problem, using convexity

properties of the $\log\|u(t)\|$.

Let us state the following

Theorem 2.1: *Let be H a Hilbert space; B a linear symmetric operator:
$D(B) \subseteq H \to H$; $u(t)$, $0 \leq t \leq T \to D(B)$, a H-continuously differentiable
function, such that $u(0) = \theta$ and $u'(t) = \gamma Bu(t)$, $o \leq t \leq T$, where γ
is any fixed complex number. Then $u(t) = \theta \; \forall t \in [o,T]$.*

Here the domain $D(B)$ is not necessarily dense in H ; B symmetric means

$(Bh,k)_H = (h,Bk)_H \; \forall h,k \in D(B)$. When $\gamma = 0$ the result is trivial. In

the general case, the result is a consequence of the following (also

interesting for itself) convexity

__Lemma:__ *Let be* $u(t)$ *,* $o \leq t \leq T \to D(B)$ *a continuously differentiable solution of the equation* $u'(t) - \gamma Bu(t) = \theta$ *. Then, if* $[a,b] \subseteq [0,T]$ *is an interval where* $\|u(t)\| > 0$ *, the scalar-valued function:* $\omega(t) = \log\|u(t)\|$ *is a convex function on* $[a,b]$ *.*

We remember that a continuous scalar function $f(t)$ on the interval $[a,b]$ is called convex when the inequality $f(\lambda t_1 + (1-\lambda)t_2) \leq \lambda f(t_1) + (1-\lambda)f(t_2)$ is satisfied, whenever $0 \leq \lambda \leq 1$, $t_1, t_2 \in [a,b]$, and that a twice continuously differentiable function $\phi(t)$ is convex on $[a,b]$ if and only if $\phi''(t) \geq 0$ $\forall t \in [a,b]$.

Let us prove now, as a preparatory step, the following

__Sublemma:__ *Under the assumptions of the Lemma and for* $\gamma \neq 0$ *, the scalar-valued function* $b(t) = (Bu(t) , u(t))$ *is continuously differentiable and its derivative equals* $2\,Re(Bu(t) , u'(t))$ *.*

__Proof:__ Let us take for example $t \in [a,b]$ and small enough h $(h > 0$ for $t = a$, $h < 0$ for $t = b)$. The differential ratio $\frac{1}{h}[b(t+h) - b(t)]$ equals $(Bu(t+h) , \frac{1}{h}[u(t+h) - u(t)]) + (\frac{1}{h}[Bu(t+h) - Bu(t)], u(t)) = (Bu(t+h),$ $\frac{1}{h}[u(t+h) - u(t)]) + (\frac{1}{h}[u(t+h) - u(t)], Bu(t))$ (by symmetry of B); now, $(Bu)(t) = \frac{1}{\gamma} u'(t)$ is continuous; it follows that $\lim\limits_{h \to o} \frac{1}{h}[b(t+h) - b(t)]$ exists and $= (Bu(t), u'(t)) + (u'(t), Bu(t)) = 2\,Re(Bu(t), u'(t))$.

We can give now the

__Proof of Lemma:__ First we remark that $\omega(t) = \frac{1}{2} \log\|u(t)\|^2$ so that it suffices to prove the convexity of the function $\phi(t) = \log(u(t), u(t))$. We shall do this by proving that $\phi(t)$ is a twice continuously differentiable function and that its second derivative $\phi''(t)$ is non-negative in $[a,b]$. Let us compute the first derivative $\phi'(t)$ which equals obviously:

$$\frac{1}{\|u(t)\|^2}\,[(u',u) + (u,u')] \quad \text{so that} \quad \phi'(t) = \frac{1}{\|u(t)\|^2}\,[(\gamma B,u) +$$

$(u,\gamma Bu)] = (2Re\gamma)\dfrac{(Bu,u)}{\|u\|^2}\,$. We apply now the above proved sublemma and find

therefore that $\phi''(t)$ is a continuous function and that

$$\frac{d^2\phi}{dt^2} = (2Re\gamma)\frac{d}{dt}\frac{(Bu,u)}{(u,u)} = 4(Re\gamma)^2\left\{\frac{\|Bu(t)\|^2}{\|u(t)\|^2} - \frac{(Bu,u)^2}{\|u(t)\|^4}\right\}\,.$$

This last expression is non-negative in virtue of the inequality

$|(Bu,u)| \leq \|Bu\|\,\|u\|$.

We can give finally the

<u>Proof of Theorem 2.1</u>: Let us assume by contradiction that $u(t)$ is not

identically null on $[o,T]$. In that case one finds at least one number

$t_o \in (0,T]$ where $\|u(t_o)\| > 0$. Let us consider now the set:

$E_{t_o} = \{t \in (0,T]$, $t \leq t_o$ and $\|u(\tau)\| > 0$ for all $\tau \in (t,t_o]\}$. Let

$\alpha = \inf E_{t_o}$. Then α is non-negative; moreover, we prove that

$\|u(\alpha)\| = 0$ whereas $\|u(t)\| > 0$ for $\alpha < t \leq t_o$.

In fact, let us assume that $\|u(\alpha)\| > 0$; using the continuity of the

scalar function $\|u(t)\|$, we can find an interval $(\alpha-\delta,\alpha+\delta)$ where still

$\|u(t)\| > 0$. On the other hand, since $\alpha = \inf E_{t_o}$, we can find an

element $\bar{t} \in E_{t_o}$, such that $\alpha \leq \bar{t} < \alpha+\delta$. Hence, $\|u(\tau)\| > 0$ for

$\tau \in (\bar{t},t_o]$. But anyhow $\|u(\tau)\| > 0$ for $\alpha-\delta < \tau < \alpha+\delta$ so that

$\|u(\tau)\| > 0$ for $\alpha-\delta < \tau \leq t_o$. By definition of E_{t_o} we obtain that

$\alpha-\delta \in E_{t_o}$ which contradicts the definition of α . Hence, we have by now

proved that: $\|u(\alpha)\| = 0$. Let us take a number $t \in (\alpha,t_o)$. Again,

there exists $t' \in E_{t_o}$, such that $\alpha \leq t' < t$. Therefore $\|u(\tau)\| > 0$

for $t' < \tau \leq t_o$, in particular for $\tau = t$. Hence, $\|u(t)\| > 0$ for

$\alpha < t < t_o$.

Let us take now a number α' , such that $\alpha < \alpha' < t_o$; on $[\alpha', t_o]$, $\|u(t)\|$ is strictly positive: we apply the Lemma on $[\alpha', t_o]$. Any point $t \in (\alpha', t_o)$ can be represented under the form: $t = \dfrac{t-\alpha'}{t_o-\alpha'}\, t_o + \dfrac{t_o-t}{t_o-\alpha'}\, \alpha'$ so that $\dfrac{t-\alpha'}{t_o-\alpha'} + \dfrac{t_o-t}{t_o-\alpha'} = 1$. Using convexity of $\log\|u(t)\|$ we find the inequality

$$-\infty < \log\|u(t)\| \leq \frac{t-\alpha'}{t_o-\alpha'}\, \log\|u(t_o)\| + \frac{t_o-t}{t_o-\alpha'}\, \log\|u(\alpha')\| \ .$$

Let us assume now that $\alpha' \downarrow \alpha$; it follows that $\|u(\alpha')\| \to 0$, the sum in the right-hand side is approaching $-\infty$ as easily seen, a contradiction that proves the Theorem.

4 Weak solutions of the nonhomogeneous equation

<u>INTRODUCTION</u>

In this Chapter we consider a (natural) class of weak = generalized solutions for nonhomogeneous first order differential equations: $u'(t) - Au(t) = f(t)$ on a real interval, in Hilbert space.   After introducing definitions and preliminary propositions, we prove, under convenient assumptions, a local (i.e., on a finite interval) existence result: given any function $f(t)$ which is Bochner square integrable on that interval, there exists at least one weak solution with the same property on the same interval (see [41], [42]).

§1.   DEFINITIONS, ELEMENTARY PROPERTIES

Let be H a Hilbert space.   Consider then an operator A which is linear, closed and densely defined in H ; hence A maps $D(A) \subseteq H$ into H , and $\overline{D(A)} = H$.   Under these conditions it is well-known that the adjoint operator A^* is also linear, closed, with dense domain $D(A^*) \subseteq H$.

For an interval $(a,b) \subset R$ which can be finite or infinite, one introduces the class of "test"-functions $K_{A^*}(a,b)$ whose members are all the functions $\varphi(t)$, $a < t < b \to D(A^*)$,   once continuously differentiable, $a < t < b \to H$, having compact support in (a,b) , such that $(A^*\varphi)(t)$ is continuous, $(a,b) \to H$.   If $a = -\infty$, $b = \infty$, we write $K_{A^*}(a,b) = K_{A^*}$.

Let us give now any function $f(t) \in B^2_{loc}(a,b;H)$ (that is, strongly measurable and such that $\int_\alpha^\beta \| f(t) \|^2_H dt < \infty$ for any compact subinterval $[\alpha,\beta] \subset (a,b)$) .   The function $u(t) \in B^2_{loc}(a,b;H)$ is said to be a generalized (or) weak solution of the equation

$$u'(t) - Au(t) = f(t) \tag{1.1}$$

on (a,b) if the integral identity

$$\int_a^b (u(t), \varphi'(t) + (A^*\varphi)(t))_H \, dt = - \int_a^b (f(t), \varphi(t))_H \, dt \tag{1.2}$$

is verified for any $\varphi \in K_{A^*}(a,b)$.

Let us consider the particular case where $f(t)$ is a strongly continuous function, $a < t < b \to H$, while $u(t)$ is H-continuously differentiable and belongs to $D(A)$ for $a < t < b$. We can prove now the very simple

Proposition 1.1: *If the equation (1.1) is verified on* (a,b) *, then the equation (1.2) is also satisfied.*

Let us take in fact a function $\varphi(t) \in K_{A^*}(a,b)$; from (1.1), taking scalar products we deduce the equality

$$(u'(t), \varphi(t))_H - (Au(t), \varphi(t))_H = (f(t), \varphi(t))_H . \tag{1.3}$$

If we integrate on (a,b), in view of the relations

$$\frac{d}{dt} (u(t), \varphi(t))_H = (u'(t), \varphi(t))_H + (u(t), \varphi'(t))_H \tag{1.4}$$

and

$$(Au(t), \varphi(t))_H = (u(t), (A^*\varphi)(t))_H , \quad \forall t \in (a,b) \tag{1.5}$$

we obtain (1.2) at once.

Another simple fact worth to be mentioned: every limit of weak solutions is again a weak solution. Precisely, one has the following

Proposition 1.2: *Let* $\{f_n(t)\}_1^\infty$ *be a sequence in* $B_{loc}^2(a,b;H)$ *which is* convergent to $f(t) \in B^2_{loc}(a,b;H)$ — i.e. $\lim\limits_{n\to\infty} \int_\alpha^\beta \| f_n(t) - f(t) \|^2_H dt = 0, \forall$ compact interval $[\alpha,\beta] \subset (a,b)$ — . Let also $\{u_n(t)\}_1^\infty$ be a sequence of weak solutions of the equation $u_n' - Au_n = f_n$, and assume that $u_n(t)$ converges to $u(t) \in B^2_{loc}(a,b;H)$. In that case $u(t)$ is again a weak solution of the equation $u' - Au = f$.

Let us write in fact the integral relation

$$\int_a^b (u_n(t), \varphi'(t) + (A^*\varphi)(t))_H dt = - \int_a^b (f_n(t), \varphi(t))_H dt$$

which is valid $\forall \varphi \in K_{A^*}(a,b)$ and $\forall n = 1,2,\dots$. If now let $n \to \infty$ one obtains the same relation for u and f instead of u_n and f_n, which proves the required proposition.

As a final remark we shall prove the

<u>Proposition 1.3</u>: Let $f(t)$ be a continuous function, $a < t < b \to H$ and assume $u(t)$, $a < t < b \to D(A)$, an H-continuously differentiable function on (a,b) such that Au is H-continuous on (a,b), which is also a weak solution of the equation $u' - Au = f$. Then $u(t)$ verifies this equation in the usual sense.

<u>Proof</u>: We see in fact, immediately, that one has

$$\int_a^b (u(t), \varphi'(t) + (A^*\varphi)(t))_H dt = - \int_a^b (u'(t) - Au(t), \varphi(t))_H dt$$

$$= - \int_a^b (f(t), \varphi(t))_H dt, \forall \varphi \in K_{A^*}(a,b),$$

which can be also written as

$$\int_a^b (u'(t) - Au(t) - f(t), \varphi(t))_H dt = 0, \forall \varphi \in K_{A^*}(a,b)$$

61

Let us take then test functions φ of a special form, namely $\varphi(t) = v(t)h$, where $v(t) \in C_o^1(a,b)$ - continuously differentiable with compact support on (a,b) - while h is an arbitrary element in $D(A^*)$. We deduce the relation

$$\int_a^b (u'(t) - Au(t) - f(t),h) \cdot v(t)dt = 0 \ , \ \ \forall h \in D(A^*) \ , \ \forall v(\cdot) \in C_o^1(a,b).$$

We shall now use the obvious continuity of the scalar-valued function $(u'(t) - Au(t) - f(t),h)$ in order to deduce vanishing of $(u' - Au - f,h)$ everywhere on (a,b) . Actually this happens for h in a dense subset of H , so that we can infer that the vector-valued function $u'(t) - Au(t) - f(t)$ equals null for all t in (a,b) .

§2. A LOCAL EXISTENCE RESULT

Under quite a mild assumption on the resolvent operator of A^* , we shall prove here the existence of (at least) one weak solution of $u' - Au = f$ on an interval (a,b) which is finite, by means of a simple "a-priori" estimate, combined with Hahn-Banach extension theorem for linear continuous functionals and with F. Riesz's representation theorem of linear continuous functionals on Hilbert spaces (see Malgrange [26] where this idea was explained in a different context - partial differential equations on bounded domains in the euclidean n-dimensional space).

Let us give now the precise statement of the

Theorem 2.1: *Let be A^* the adjoint operator to the linear, closed, densely defined operator A in the Hilbert space H . Let us assume that the resolvent operator $(\lambda I - A^*)^{-1}$ exists and belongs to $L(H;H)$ on a vertical line: $\lambda = \sigma_o + i\tau$, $-\infty < \tau < \infty$ in the complex plane, and that the uniform bound $\| (\lambda I - A^*)^{-1} \|_{L(H;H)} \leq L$, $-\infty < \tau < \infty$ is satisfied. Then, given any finite interval $(a,b) \subset R$ and any function $f \in B^2(a,b;H)$, there* *exists at least one function* $u(t) \in B^2(a,b;H)$ *such that (1.2) is verified,*

$\forall \varphi \in K_{A*}(a,b)$.

We start the proof of this result with an a priori estimate given in the

form of

Lemma 2.1: *Under the assumptions of Theorem 2.1, there exists a constant*

$C_{a,b} > 0$ *in such a way that the estimate*

$$(\int_a^b \|\psi(t)\|_H^2 dt)^{\frac{1}{2}} \le C_{a,b} (\int_a^b \|\psi'(t) + A^*\psi(t)\|_H^2 dt)^{\frac{1}{2}} \qquad (2.1)$$

is satisfied, $\forall \psi \in K_{A*}(a,b)$.

Proof: Let us introduce the H-valued function $\varphi(t) = e^{\sigma_o t}\psi(t)$; if we

denote $\psi'(t) + A^*\psi(t) = \phi(t)$ we can see immediately that

$$\varphi'(t) + (A^* - \sigma_o I)\varphi(t) = \phi(t)e^{\sigma_o t} = g(t) \qquad (2.2)$$

where $\varphi(t)$ and $g(t)$ have both compact support in (a,b) . Let us consider

now the inverse Fourier transforms

$$\overset{\vee}{\varphi}(\tau) = \frac{1}{\sqrt{2\pi}} \int_{-\infty}^{\infty} e^{i\tau t}\varphi(t)dt \, , \quad \overset{\vee}{g}(\tau) = \frac{1}{\sqrt{2\pi}} \int_{-\infty}^{\infty} e^{i\tau t}g(t)dt \, . \qquad (2.3)$$

It is easy to prove, following the scalar-case pattern, the so called

Parseval's identity; precisely, the equalities

$$\int_{-\infty}^{\infty} \|\overset{\vee}{\varphi}(\tau)\|_H^2 d\tau = \int_{-\infty}^{\infty} \|\varphi(t)\|_H^2 dt \, , \quad \int_{-\infty}^{\infty} \|\overset{\vee}{g}(\tau)\|_H^2 d\tau = \int_{-\infty}^{\infty} \|g(t)\|_H^2 dt \, . \qquad (2.4)$$

Now, let us multiply both sides in (2.2) by $\exp(i\tau t)$, and then integrate

on R : we deduce the relation

$$\frac{1}{\sqrt{2\pi}} \int_{-\infty}^{\infty} \overset{\vee}{\varphi}{}'(t) e^{i\tau t} dt + \frac{1}{\sqrt{2\pi}} \int_{-\infty}^{\infty} (A^* - \sigma_o I) e^{i\tau t} \overset{\vee}{\varphi}(t) dt = \frac{1}{\sqrt{2\pi}} \int_{-\infty}^{\infty} e^{i\tau t} \overset{\vee}{g}(t) dt$$

$$(2.5)$$

and therefore, through partial integration, making use of closedness of A^*, we see that $\overset{\vee}{\varphi}(\tau) \in D(A^*)$ and we obtain the equality

$$[A^* - (\sigma_o + i\tau)I]\overset{\vee}{\varphi}(\tau) = \overset{\vee}{g}(\tau) , \quad -\infty < \tau < \infty .$$

$$(2.6)$$

Using our assumption in Theorem 2.1 one finds the representation formula

$$\overset{\vee}{\varphi}(\tau) = -[(\sigma_o + i\tau)I - A^*]^{-1}\overset{\vee}{g}(\tau) , \quad -\infty < \tau < \infty$$

$$(2.7)$$

and, accordingly, the uniform bound

$$\| \overset{\vee}{\varphi}(\tau) \|_H \leq L \| \overset{\vee}{g}(\tau) \|_H , \quad -\infty < \tau < \infty .$$

$$(2.8)$$

Consequently, one sees that

$$\int_{-\infty}^{\infty} \| \overset{\vee}{\varphi}(t) \|_H^2 dt = \int_{-\infty}^{\infty} e^{2\sigma_o t} \| \psi(t) \|_H^2 dt \leq L^2 \int_{-\infty}^{\infty} \| \overset{\vee}{g}(t) \|_H^2 dt = L^2 \int_{-\infty}^{\infty} e^{2\sigma_o t} \| \phi(t) \|_H^2 dt$$

$$(2.9)$$

or also

$$\int_a^b e^{2\sigma_o t} \| \psi(t) \|_H^2 dt \leq L^2 \int_a^b e^{2\sigma_o t} \| \psi'(t) + A^*\psi(t) \|_H^2 dt .$$

$$(2.10)$$

If, for example, σ_o is negative, one infers from (2.10) the inequality

$$e^{2\sigma_o b} \int_a^b \| \psi(t) \|_H^2 dt \leq L^2 e^{2\sigma_o a} \int_a^b \| \psi'(t) + A^*\psi(t) \|_H^2 dt$$

$$(2.11)$$

which gives the Lemma with $C_{a,b} = L e^{\sigma_o(a-b)}$, while for $\sigma_o > 0$ we get

$$C_{a,b} = Le^{\sigma_o (b-a)} \qquad \text{so that in any case} \quad C_{a,b} = Le^{|\sigma_o|(b-a)} \, .$$

Proof of Theorem 2.1: Let us introduce the subset M of the space $B^2(a,b;H)$ which is the range of the operator $\frac{d}{dt} + A^*$ defined on $K_{A^*}(a,b)$:

$M = \{\psi' + A^*\psi \, , \ \psi \in K_{A^*}(a,b)\}$; it is therefore a linear subset of $B^2(a,b;H)$.

Consider now the mapping F of M in the complex plane which is defined through the equality

$$F(\varphi' + A^*\varphi) = - \int_a^b (f(t),\varphi(t))_H dt \tag{2.12}$$

where $f \in B^2(a,b;H)$ is the one assumed in Theorem 2.1.

This map is well-defined (as a single-valued function), since, according to the above Lemma, if $\varphi_1' + A^*\varphi_1 = \varphi_2' + A^*\varphi_2$, then $\varphi_1 = \varphi_2$. Furthermore we see that F is an anti-linear map:

Let $\phi \in M$, so that $\phi = \varphi' + A^*\varphi$, $\varphi \in K_{A^*}(a,b)$; then $\lambda\phi = (\lambda\varphi)' + A^*(\lambda\phi)$ and $F(\lambda\phi) = - \int_a^b (f(t), (\lambda\phi)(t))_H dt = -\bar{\lambda} \int_a^b (f(t),\varphi(t))_H dt = \bar{\lambda}F(\phi)$. Also, $F(\phi_1 + \phi_2) = F(\phi_1) + F(\phi_2)$, $\forall \phi_1,\phi_2 \in M$.

Moreover, F is a continuous map on M considered as subspace of $B^2(a,b;H)$. One has in fact the estimate, for $\phi = \varphi' + A^*\varphi$, $\varphi \in K_{A^*}(a,b)$

$$|F(\phi)| = \left| \int_a^b (f(t),\varphi(t))_H dt \right| \ \leq \ \left(\int_a^b \| f(t) \|_H^2 dt\right)^{\frac{1}{2}} \left(\int_a^b \| \varphi(t) \|_H^2 dt\right)^{\frac{1}{2}}$$

$$\tag{2.13}$$

whose consequence, in view of (2.1), is the inequality

$$|F(\phi)| \ \leq \ \left(\int_a^b \| f(t) \|_H^2 dt\right)^{\frac{1}{2}} C_{a,b} \left(\int_a^b \| \phi(t) \|_H^2 dt\right)^{\frac{1}{2}} , \quad \forall \phi \in M \ . \tag{2.14}$$

Let us extend now, according to well-known results, the mapping F from M to the whole space $B^2(a,b;H)$, as an anti-linear continuous mapping $\tilde{F}$ of $B^2(a,b;H)$ into the complex plane, such that $\tilde{F} = F$ on M . According to

Riesz's representation theorem, one can find a function $u(t) \in B^2(a,b;H)$, in such a way that $\tilde{F}(v) = \int_a^b (u(t),v(t))_H dt$, $\forall v \in B^2(a,b;H)$; in particular, when $v \in M$, $v = \varphi' + A^*\varphi$ where $\varphi \in K_{A^*}(a,b)$ and $\tilde{F}(v) = F(v) = F(\varphi'+A^*\varphi)$
$$= -\int_a^b (f(t),\varphi(t))_H dt = \int_a^b (u(t),\varphi'(t) + A^*\varphi(t))_H dt \ .$$
Here φ can be any member of $K_{A^*}(a,b)$, so that the $u(\cdot) \in B^2(a,b;H)$ which is granted by Riesz's theorem is one possible weak solution of the equation $u' - Au = f$. This ends the proof of Theorem (2.1).

Remark: If the homogeneous equation $u' - Au = \theta$ has non-trivial (weak) solutions (this happens "often" but not always, see [2]) then the solution u of $u' - Au = f$ is certainly not the only one.

5 Asymptotic behavior

<u>INTRODUCTION</u>

In this Chapter we study certain results concerning the behavior of solutions of a differential equation $u'(t) = Au(t) + f(t)$, A being a linear, unbounded operator in a Hilbert or a Banach space, when the "time" variable t runs over the whole real line, or over the positive half-line. We can say that we examine behavior "at infinity" or "asymptotic" of the solutions. We shall consider here some results concerning uniqueness of bounded solutions over the real line, but no results on existence of such bounded solutions; some other results concerning solutions which have a (definite) limit at the infinity and also solutions whose norm grows indefinitely when $t \to \infty$.

<u>§1.</u>   <u>UNIQUENESS OF BOUNDED SOLUTIONS</u>

The first result in this section concerns those solutions of a homogeneous differential equation: $u'(t) = Au(t)$, which are defined on the whole real line, the underlying space being a Banach space. We shall state and prove the following[1]

<u>Theorem 1.1</u>: *Let be X a Banach space and let T_t, $t \geq 0 \to L(X,X)$ be a one-parameter semi-group of class C_o, verifying an inequality:*

$\| T_t \|_{L(X,X)} \leq Me^{\beta t}$, $t \geq 0$, *where $M > 0$ and $\beta < 0$. Let be A the infinitesimal generator of T_t, and let $u(t)$, $-\infty < t < \infty \to D(A)$ be a strong solution of the equation: $u'(t) - Au(t) = \theta$, $t \in R$. Then, if*

---

[1] See reference [46].

67

one has

$$\sup_{t \in R} \int_{t}^{t+1} \| u(\sigma) \|^2 d\sigma < \infty$$

it results that $u(t) = \theta$, $t \in R$.

(An equivalent formulation is the following: if $u(t)$ is a solution which is not identically zero of the equation $u' = Au$, $t \in R$, then:

$$\sup_{t \in R} \int_{t}^{t+1} \| u(\sigma) \|^2 d\sigma = +\infty.)$$

<u>Remark 1</u>: A function $v(t)$, $t \in R \rightarrow X$ is said to be bounded on R if $\sup_{t \in R} \| v(t) \| < \infty$, and is said to be S^2-bounded if

$$\sup_{t \in R} \int_{t}^{t+1} \| v(\sigma) \|^2 d\sigma < \infty \quad ;$$

the first concept of boundedness is obviously stronger than the second one, so that in Theorem 1.1 we prove a bit more than uniqueness of bounded solutions, namely, uniqueness of S^2-bounded solutions.

<u>Remark 2</u>: In view of Theorem 7.1 (Chapter II), any linear closed densely defined operator A in X , such that for $\lambda > \beta$, $(\lambda - A)^{-1} \in L(X,X)$ and $\| (\lambda - A)^{-n} \| \leq M(\lambda - \beta)^{-n}$, $n = 1, 2, \ldots$ is verified, is the infinitesimal generator of a semi-group as requested in this Theorem.

In order to demonstrate our Theorem, we need a result of a general character, establishing a representation formula for solutions on the whole real line of equations $v' = Av$ where A is generator of a semi-group of class C_o . Let us give it in the form of the following

68

Lemma 1: Let U_t , $t \geq 0 \to L(X,X)$ be an operator semi-group of class C_o
and let B its infinitesimal generator.   Assume that $v(t)$, $t \in R \to D(B)$
is an X-continuously differentiable solution of the equation $v' = Bv$.
Then, for any fixed real number a and for all $t \geq a$, the representation
formula

$$v(t) = U_{t-a}v(a)$$

is valid.

Proof of Lemma: Let us effectuate the substitution $t = a + \sigma$, $t \geq a$,
and hence $v(t) = v(a + \sigma) = w(\sigma)$, $\sigma \geq 0$.   We see that $w(\sigma)$ is
continuously differentiable, $\sigma \geq 0 \to X$ and belongs to $D(B)$ $\forall$ such σ ;
moreover: $w'(\sigma) = v'(a+\sigma) = Bv(a+\sigma) = Bw(\sigma)$, $\sigma \geq 0$.   On the other hand,
the function $z(\sigma) = U_\sigma w(0)$ verifies, for $\sigma \geq 0$, the same equation:
$z'(\sigma) = Bz(\sigma)$; also, $z(0) = w(0)$.   By the uniqueness of the initial-
value problem established in Theorem 4.2 (Chapter II) we infer that
$w(\sigma) = z(\sigma)$, $\forall \sigma \geq 0$, that is $v(a + \sigma) = U_\sigma v(a)$, $\sigma \geq 0$ which translates
as $v(t) = U_{t-a}v(a)$, $t \geq a$, in the former variable t .

Going on with the proof of our Theorem let us state the following

Lemma 2: Let $z(t)$, $t \in R \to X$ be a continuous function which is S^2-bounded:
$$\sup_{t \in R} \int_t^{t+1} \| z(t) \|^2 dt < \infty .$$ Then, there exists a sequence of real numbers
$(t_n)_1^\infty$ such that $\lim_{n \to \infty} t_n = -\infty$, and $\sup_{n \in N} \| z(t_n) \|_X \leq L < \infty$.

Proof of Lemma 2: Let us denote by L^2 the number $\sup_{t \in R} \int_t^{t+1} \| z(t) \|^2 dt$;
it follows, $\forall n \in N$,

$$\int_{-n}^{-n+1} \| z(\sigma) \|^2 d\sigma \leq L^2 .$$

Remark also that the scalar-valued function $\phi(\sigma) = \|z(\sigma)\|^2$ is a continuous one; we can apply the mean-value theorem for integrals and find, $\forall n \in N$, a number $t_n \in [-n, -n+1]$, in such a way that

$$\int_{-n}^{-n+1} \|z(\sigma)\|^2 d\sigma = \|z(t_n)\|^2, \quad \text{so that} \quad \|z(t_n)\|^2 \le L^2, \ n \in N.$$

We are now ready to finish the

<u>Proof of Theorem 1.1</u>: Let $u(t)$ be the S^2-bounded solution of the equation $u' = Au$ on the real line; let us fix a real number t; using the above found sequence $(t_n) \to -\infty$ we find, for $n \ge n_o(t)$, that $t_n < t$.

This permits use of representation formula from Lemma 1, so that $u(t) = T_{t-t_n} u(t_n)$ for all $n \ge n_o(t)$, and this permits the estimate:

$$\|u(t)\| < \|T_{t-t_n}\|_{L(X,X)} \cdot \|u(t_n)\| \le Me^{\beta(t-t_n)} \cdot L, \ \forall n \ge n_o(t).$$

If we let now n to increase beyond any limit, $t - t_n$ will tend to $+\infty$, so that $MLe^{\beta(t-t_n)} \to 0$. The left hand side, $\|u(t)\|$, is independent of n, so that $\|u(t)\| = 0$. This proves the Theorem.

<u>Remark 1</u>: If A is the infinitesimal generator of a C_o-semi-group verifying the above estimate: $\|T_t\| \le Me^{\beta t}$, $\beta < 0$, $\forall t \ge 0$, and if one considers the non-homogeneous equation: $u'(t) - Au(t) = f(t)$ over the real line, it is quite easy to prove, under the assumption that $f(t)$ is bounded over R, the existence of the (unique) bounded over R solution (in strong or in weaker sense) $u(t)$ of this equation. However, as stated in the Introduction, we shall not go further in investigation of this new problem.

<u>Remark 2</u>: The above assumption on the semi-group T_t (ensuring exponential decay at $+\infty$ for the operator norm of this semi-group) is quite a restrictive one, even if it is quite common in many important particular cases. It is

70

therefore of interest to consider semi-groups which tend to θ as $t \to +\infty$ in somewhat "weakened" sense, for example if $\lim\limits_{t \to \infty} \| T_t x \|_X = 0 \ \forall x \ \varepsilon \ X$, which also can be often found to hold in some (other) special cases. In this new situation we are, by now, unable to produce existence theorems for bounded solutions of the nonhomogeneous equation which was discussed above, but we are still able to give results concerning uniqueness of bounded solutions (if any at all!). Let us state and prove now a result on this line, precisely the

Theorem 1.2[1] *Let $T(t)$ be a C_o-semi-group in the Banach space X and let us assume one of the hypothesis (α) or (β) below:*

(α) $\lim\limits_{t \to \infty} T_t x = \theta$ *,* $\forall x \ \varepsilon \ X$ *and* $(\lambda_o - A)^{-1}$ *is a compact operator in X for at least one complex number λ_o, A being the infinitesimal generator to T_t .*

(β) $\lim\limits_{t \to \infty} T_t^* x^* = \theta \ \forall x^* \ \varepsilon \ X^*$ *(dual space of X) , where $T^*(t)$ is, for any $t \geq 0$, the dual operator to $T(t)$, acting in X^* .*

Let $u(t)$ be a solution not identically zero of the equation $u'(t) = Au(t)$ over the whole real line. Then: $\sup\limits_{t \varepsilon R} \int\limits_{t}^{t+1} \| u(\sigma) \|^2 d\sigma = +\infty$ *.*

Proof: In order to demonstrate part (α) we need the following preliminary

Lemma: *The semi-group T_t commutes with the resolvent operator $(\lambda_o - A)^{-1}$*
We shall use a representation formula established previously, precisely:

$$(\lambda_o - A)^{-1} x = \int\limits_{o}^{\infty} e^{-\lambda_o t} T_t x \, dt \ , \ \forall x \ \varepsilon \ X$$

---

[1] See reference [44].

one deduces:

$$(\lambda_o - A)^{-1} T_\sigma x = \int_o^\infty e^{-\lambda_o t} T_t T_\sigma x\, dt = \int_o^\infty T_\sigma (e^{-\lambda_o t} T_t x)\, dt = T_\sigma (\lambda_o - A)^{-1} x \ ,$$

$$\forall x \ \varepsilon \ X \ , \quad \forall \sigma \geq 0 \ ,$$

as required.

Continuing now the proof of the Theorem we shall define, along with a function $u(t)$, solution of $u' - Au = \theta$ and verifying:
$$\sup_{t\varepsilon R} \ \int_t^{t+1} \| u(\sigma) \|^2 d\sigma = L^2 < \infty \ , \quad \text{the "regularized" function}$$
$w(t) = (\lambda_o - A)^{-1} u(t)$. We shall prove below that $w(t) = \theta \ \forall t \ \varepsilon \ R$ and this will imply that $u(t) = \theta$, $\forall t \ \varepsilon \ R$.

First we shall apply to $u(t)$ the Lemma 2 in previous Theorem, so that we find a sequence of real numbers $(t_n) \to -\infty$ such that $\| u(t_n) \| \leq L$; the sequence $(u(t_n))_1^\infty \subseteq X$ is bounded; its image $(w(t_n))_1^\infty = \{(\lambda_o - A)^{-1} u(t_n)\}_1^\infty$ under the compact operator $(\lambda_o - A)^{-1}$ will be a relatively compact sequence in X; a subsequence of it, say $\{w(t_{n_k})\}_{k=1}^\infty$ will be (strongly) convergent in X towards an element $w_\infty \ \varepsilon \ X$.

Let us proceed now as in the proof of Theorem 1.1; given any fixed real t we find $k_o(t)$ such that $t_{n_k} < t$ for $k \geq k_o$. We can represent $u(t)$ for $t > t_{n_k}$ by the formula: $u(t) = T_{t-t_{n_k}} u(t_{n_k})$, and if we apply at both sides the operator $(\lambda_o - A)^{-1}$ and use also the commutativity of this operator with the semi-group T_t, we obtain $(\lambda_o - A)^{-1} u(t) = T_{t-t_{n_k}} (\lambda_o - A)^{-1} u(t_{n_k})$ $k \geq k_o(t)$, hence

$$w(t) = T_{t-t_{n_k}} w(t_{n_k}) \ , \ k \geq k_o(t) \ .$$

We shall now write $w(t)$ under the form

$$w(t) = T_{t-t_{n_k}} [w(t_{n_k}) - w_\infty] + T_{t-t_{n_k}} w_\infty$$

and we shall also remark that the assumption: $\lim_{t \to \infty} T_t x = \theta$, $\forall x \in X$ will

imply, via the uniform boundedness theorem, **a** uniform estimate:

$\| T_t \| \leq A$, $t \geq 0$. This permits us to derive the inequality

$$\| w(t) \| \leq A \| w(t_{n_k}) - w_\infty \| + \| T_{t-t_{n_k}} w_\infty \| , \quad k \geq k_o(t) .$$

If $k \to \infty$, the right-hand side tends to 0 ; hence $w(t) = \theta$, $\forall t \in R$, which proves (α) .

The proof for the part (β) is somehow simpler but quite similar; again, under the assumption that $\sup_{t \in R} \int_t^{t+1} \| u(\sigma) \|^2 d\sigma < \infty$ it is possible to find a sequence $(t_n)_1^\infty$ convergent to $-\infty$, such that the sequence $\{u(t_n)\}_1^\infty$ be bounded in X . Fix once more a real t ; then $t_n \leq t$ for $n \geq N(t)$ and we can write: $u(t) = T_{t-t_n} u(t_n)$. Let us take now an arbitrary element in the dual space of X , i.e. a linear continuous functional $x^* \in X^*$. We can write therefore the equality((where $< , >$ means the duality between X and X^*):

$$<x^*, u(t)> = <x^*, T_{t-t_n} u(t_n)> = <T^*_{t-t_n} x^* , u(t_n)>$$

and, accordingly, the inequality

$$|<x^*, u(t)>| \leq \| T^*_{t-t_n} x^* \| \ \| u(t_n) \| \leq (\sup_{n \in N} \| u(t_n) \|) \| T^*_{t-t_n} x^* \| .$$

If $n \to \infty$, using the assumption (β) we obtain that the right-hand side is

convergent to 0 . The left hand side being independent of n , it follows that $\langle x^*, u(t)\rangle = 0$.   This is true for *any* element x^* in the dual space X^* .   By a classical Corollary of the Hahn-Banach theorem (see for example [16] - pg. 33, Theorem 2.8.2) we obtain that $u(t) = \theta$.   This is true for any real t and $u(t)$ is the null solution, a contradiction which proves also part *(β)*.

Continuing our "series" of "uniqueness of bounded solutions" theorems, we shall restrict now ourselves to a Hilbert space (instead of a general Banach space) thus permitting greater generality in the choice of the operator A which is not necessarily generator of a C_o-operator semi-group.   The nice (and simple) result below was proved by H. Levine [23], thus extending a previous result of the author concerning the self-adjoint case [45]:

<u>Theorem 1.3</u>: *Let be A a linear symmetric operator with domain $D(A)$ in the Hilbert space H , such that $Ax = \theta$ iff $x = \theta$.   Let be $u(t)$, $-\infty < t < \infty \rightarrow D(A)$, a H-continuously differentiable solution of the equation $u'(t) = Au(t)$, $t \in R$, and let us assume that:* $\displaystyle \sup_{t\in R} \int_t^{t+1} \|u(\sigma)\|^2 d\sigma < \infty$. *Then $u(t) = \theta$, $t \in R$.*

As we did in Chapter III, §2 where we proved uniqueness of Cauchy problem using convexity methods, here again the proof will be based on some simple properties of convex functions.

We shall denote by $F(t)$ the function $\displaystyle \int_t^{t+1} \|u(\eta)\|^2 d\eta$; the derivative $F'(t)$ exists and equals $\|u(t+1)\|^2 - \|u(t)\|^2$ which in turn equals $\displaystyle \int_t^{t+1} \left(\frac{d}{d\eta} \|u(\eta)\|^2\right) d\eta$.   (The square of the norm of u equals $(u(\eta), u(\eta))$ and is therefore continuously differentiable; its derivative equals $(u'(\eta), u(\eta)) + (u(\eta), u'(\eta)) = 2 Re(u'(\eta), u(\eta))$.)

Accordingly we are entitled to write the equality:

$$F'(t) = 2 \int_{t}^{t+1} Re(u'(\eta), u(\eta))d\eta = 2 \int_{t}^{t+1} Re(Au(\eta), u(\eta))d\eta =$$

$$2 \int_{t}^{t+1} (Au(\eta), u(\eta))d\eta$$

(using symmetry of A).

The integrand here is obviously a continuous function; hence we can deduce the existence of the second derivative $F''(t)$ which equals $2(Au(t+1),u(t+1)) - 2(Au(t),u(t))$.

Let us remember now the *Sublemma* in Chapter II, §2; thus, the scalar function: $(Au(\eta), u(\eta))$ is continuously differentiable; its derivative equals $2 Re(Au(\eta), u'(\eta)) = 2\|Au(\eta)\|^2$.

Therefore, we may write again

$$F''(t) = 2 \int_{t}^{t+1} \frac{d}{d\eta} (Au(\eta),u(\eta))d\eta = 4 \int_{t}^{t+1} \|Au(\eta)\|^2 d\eta , \quad \forall t \in R$$

and observe that $F''(t) \geq 0 \ \forall t \in R$ (hence $F(t)$ is a convex function of t). Our assumption in the theorem is that $F(t)$ is a bounded function on the real line. Let us assume also, by contradiction, that $\|u(t_o)\| > 0$ for some $t_o \in R$. Therefore, according to another of the assumptions, $\|Au(t_o)\|$ is also strictly positive. This implies: $\|Au(\sigma)\| > 0$ in a neighbourhood of t_o, and in turn it follows that $F(t)$ is non-constant on R-otherwise $F''(t) = 4 \int_{t}^{t+1} \|Au(\eta)\|^2 d\eta = 0 , \ \forall t , \Rightarrow \|Au(\eta)\| = 0 , \ \forall \eta \in R.$

Let us end the proof; we show the impossibility that a non-constant twice differentiable convex function on R be also bounded on R. In fact: if $F(t)$ is not constant, there is $t^* \in R$ such that, say, $F'(t^*)$ is strictly

< 0 ; for $t < t^*$, we see that $F'(t) \leq F'(t^*)$; use the mean-value theorem: $F(t) - F(t^*) = (t-t^*)F'(\xi)$, $t < \xi < t^*$; here $t - t^* < 0$, $F'(\xi) \leq F'(t^*)$, hence $F(t) \geq F(t^*) + (t-t^*)F'(t^*)$; if $t \to -\infty$ $(t-t^*)F'(t^*) \to +\infty$, and $F(t)$ grows indefinitely, a contradiction.

This proves Theorem 1.3.

§2.   SOLUTIONS WHICH HAVE A LIMIT AT INFINITY

Our first result here is stated in the following

Theorem 2.1: _Let_ H _be a Hilbert space;_ A ; $D(A) \subseteq H \to H$, _a linear closed operator which is densely defined;_ _assume furthermore that_

$$Re(Ah,h) \leq \beta(h,h) , h \in D(A) , Re(A^*k,k) \leq \beta(k,k) , k \in D(A^*) .$$

$(A^*$ _is the adjoint operator to_ A) _where_ β _is a negative number._ _Let_ $u(t)$, $t \geq 0 \to D(A)$ _be a solution of the nonhomogeneous equation:_ $u'(t) = Au(t) + f_o$, _where_ f_o _is given in_ H . _Then, there exists_ $w_o \in H$ _such that:_ $\lim\limits_{t\to\infty} u(t) = w_o$.

Proof: First, we refer to the Theorem 8.1 in §8 of Chapter II; the sequence of estimates: $\| (\lambda-A)^{-n} \| \leq \dfrac{1}{(\lambda-\beta)^n}$, $\forall \lambda > \beta$ is satisfied, hence an application of Theorem 7.2 + Remark in Chapter II indicates that A is the infinitesimal generator of a C_o-semi-group T_t which verifies the estimate $\| T_t \| \leq e^{\beta t}$, $\forall t \geq 0$. From Theorem 4.2 (Chapter II) we deduce also that any solution of the homogeneous equation: $v' = Av$, $t \geq 0$, $v(0) = v_o \in D(A)$ is given by the formula: $v(t) = T_t v(0)$.

On the other hand, we see that the inverse operator A^{-1} exists and belongs to $L(H;H)$; let be $w_o = -A^{-1} f_o$; $u(t)$ a solution of $u' = Au + f_o$; $v(t) = u(t) - w_o$; then $v(t)$ is a continuously differentiable solution of

76

the equation: $v' = u' = Au + f_o = A(v(t) + w_o) + f_o = Av(t)$; hence, by the

above, $v(t) = T_t v(0)$, $\|v(t)\| \leq e^{\beta t}\|v(0)\| \to 0$ as $t \to \infty$; thus:

$\lim_{t\to\infty} \|u(t) - w_o\| = 0$, so that $u(t) \to -A^{-1}f_o$ when $t \to \infty$

Remark: The result just proved applies in particular to self-adjoint

operators $(A = A^*)$ in Hilbert spaces, which are negative definite. The case

where A self-adjoint is ≤ 0 without being negative definite (so that

$(Ah,h) \leq 0$, $\forall h \in D(A)$) is a more complex one, and the solutions of the

nonhomogeneous equation: $u'(t) = Au(t) + f_o$, on the positive half-line, may

behave in quite different ways. The simplest result is the following

Theorem 2.2: *Let $A \leq 0$ be a self-adjoint operator in the Hilbert space H,
and assume that there exists $f_o \in D(A)$, $f_o \neq \theta$, and $Af_o = \theta$. Then for
any solution $u(t)$ of the equation: $u'(t) = Au(t) + f_o$ we have:*

$\lim_{t\to\infty} \|u(t)\| = \infty$.

Proof: In fact, again by Theorems 8.1 and 7.2 + Remark in Chapter II, we see

that A is the infinitesimal generator of an operator semi-group T_t of

class C_o where $\|T_t\| \leq 1$, $\forall t \geq 0$ (contraction semi-group). Now, any

solution $u(t)$ of $u' = Au + f_o$ can be represented under the form:

$u(t) = T_t u(0) + tf_o$. In fact, we see that: $\frac{d}{dt}(T_t u(0) + tf) = AT_t u(0) + f_o$.

Hence, $v(t) = T_t u(0) + tf_o$ is a solution to: $v'(t) = AT_t u(0) + f_o =$

$A(T_t u(0) + tf_o) + f_o$ (because $Af_o = \theta$), hence $v'(t) = Av(t) + f_o$;

moreover, $v(0) = u(0)$; therefore (Theorem 4.2, Chapter II) , $v(t) = u(t)$,

$\forall t \geq 0$.

Once we know that $u(t)$ equals necessarily $T_t u(o) + tf_o$, we estimate

it from below by: $\|u(t)\| = \|T_t u(0) + tf_o\| \geq \|tf_o\| - \|T_t u(0)\| \geq$

$t\|f_o\| - \|u(0)\|$ (using also $\|T_t\| \leq 1$) .   As we assumed $\|f_o\| > 0$, the required conclusion follows.

The next possibility for a self-adjoint operator A which is negative without being negative definite arises when $\lambda = 0$ is *not* an eigen-value of the operator (as it was in Theorem 2.2), which can be expressed by: $Ah = \theta$ iff $h = \theta$.   In this case we shall explain two results: we shall see that the solutions of the nonhomogeneous equation: $v' = Av + f_o$ behave quite differently: when f_o belongs to the range of A - hence there exists $f_1 \in D(A)$ such that $Af_1 - f_o -$, all solutions $v(t)$ approach $-f_1$ as $t \to \infty$, as it happens in Theorem 2.1; on the other hand, as was indicated to us by Professor J. J. Schäffer some years ago in an exchange of correspondence, when f_o *does not* belong to the range of A , all solutions of $v' = Av + f_o$ have a norm increasing to ∞ for $t \to \infty$, similarly to Theorem 2.2.   Precise statements follow:

Theorem 2.3: *Let $A \leq \theta$ be a self-adjoint operator with domain $D(A)$ in the Hilbert space H, such that $Ah = \theta$ iff $h = \theta$.   Let $f_o = Af_1$, where $f_1 \in D(A)$.   Then the function $v(t) = -f_1$, $0 \leq t < \infty$ is a solution of the equation $v' = Av + f_o$ and any other solution $w(t)$ of this equation has the property that: $\lim\limits_{t \to \infty} w(t) = -f_1$.*

Proof: $v'(t) = \theta$ $\forall t \geq 0$; $Av + f_o = -Af_1 + f_o = \theta$, $\forall t \geq 0$;   therefore $v' = Av + f_o$, $\forall t \geq 0$.   If $w(t)$ is any solution of $w' = Aw + f_o$, then $(w-v)' = A(w-v)$; hence $(w-v)(t) = T_t(w-v)(0)$; $w(t) = v(t) + T_t w(0) - T_t v(0) = -f_1 + T_t(w(0) + f_1)$.   In order to show that $\lim\limits_{t \to \infty} w(t) = -f_1$ it will suffice to prove (see our paper [45]) the following

<u>Lemma</u>: *If $\lambda = 0$ is not an eigen-value for the self-adjoint operator $A \leq \Theta$ then, for the semi-group T_t generated by A holds:* $\lim\limits_{t\to\infty} T_t h = \theta$, $\forall h \in H$.

In order to prove this result we shall use von Neumann's spectral theory for self-adjoint operator in Hilbert spaces (see [34], [37]).   There exists therefore a spectral family of projection operators in H , $\{E_\lambda\}_{-\infty}^{\infty}$, such that $\lim\limits_{\lambda\to-\infty} E_\lambda x = \theta$, $\forall x \in H$, $\lim\limits_{\lambda\to+\infty} E_\lambda x = x$, $\forall x \in H$, $E_\lambda E_\mu = E_{\min(\lambda,\mu)}$, $\lim\limits_{\mu\downarrow\lambda} E_\mu x = E_{\lambda+0} x = E_\lambda x$, $\forall x \in H$, $E_\lambda = I$ for $\lambda \geq 0$ (because $A \leq \Theta)$, and $A = \int\limits_{-\infty}^{0} \lambda dE_\lambda$;   this integral exists for example in the sense that,

$$(Ah,k) = \int\limits_{-\infty}^{0} \lambda d(E_\lambda h,k) , \quad \forall h \in D(A) , \quad \forall k \in H \text{ and } D(A) = \{x \in H , \text{ such that } \int\limits_{-\infty}^{0} |\lambda|^2 d\| E_\lambda x\|^2 < \infty\} .$$

We can form the function of A , $\exp(tA)$ corresponding to the real-valued continuous function $\exp(t\lambda)$, for $t \geq 0$.   Therefore:
$\exp(tA)x = \int\limits_{-\infty}^{0} e^{\lambda t} dE_\lambda x$ and obviously we have

$$\| \exp(tA)x\|^2 = \int\limits_{-\infty}^{0} e^{2\lambda t} d\| E_\lambda x\|^2 \leq \int\limits_{-\infty}^{0} d\| E_\lambda x\|^2 = \| x\|^2$$

so that $\| \exp(tA)\| \leq 1$.

We shall now prove, under the assumption $" Ax = \theta$ iff $x = \theta"$, that $\lim\limits_{t\to\infty} \| \exp(tA)x\| = 0$, $\forall x \in H$ and the demonstration will be ended when we shall also prove that $T_t = \exp(tA)$.   Let us write therefore the equality

$$\| \exp(tA)x\|^2 = \int\limits_{-\infty}^{0} e^{2\lambda t} d\| E_\lambda x\|^2 = \int\limits_{-\infty}^{\delta} e^{2\lambda t} d(E_\lambda x,x) + \int\limits_{\delta}^{0} e^{2\lambda t} d(E_\lambda x,x)$$

where δ is an arbitrary negative number.   We deduce, because of inequalities:

$e^{2\lambda t} \leq e^{2\delta t}$ for $\lambda \leq \delta$, $t \geq 0$, and $e^{2\lambda t} \leq 1$, $\delta \leq \lambda \leq 0$, $t \geq 0$, the estimate $\| \exp(tA)x \|^2 \leq e^{2\delta t} \| x \|^2 + (\| x \|^2 - \| E_\delta x \|^2)$.

Now, we shall use the well-known fact (see e.g. [1]) that the spectral family $\{E_\lambda\}_{-\infty}^{\infty}$ of a self-adjoint operator is strongly continuous (right *and* left) in any point $\lambda \in R$ which is not an eigen-value for A ; thus we see that $\lim_{\delta\uparrow 0} \| E_\delta x \|^2 = \| x \|^2$.

Let us take now $\varepsilon > 0$; then, we first choose (and fix) $\delta(\varepsilon)$ in order to have $\| x \|^2 - \| E_\delta x \|^2 < \frac{\varepsilon^2}{2}$; next, we can find $\bar{t}$ depending on $\delta(\varepsilon)$ and on x , such that $e^{2\delta t} \| x \|^2 < \frac{\varepsilon^2}{2}$ for $t \geq \bar{t}$. Accordingly, we obtain $\| \exp(tA)x \|^2 \leq \varepsilon^2$ for $t \geq \bar{t}(\varepsilon,x)$ which proves the strong convergence to θ of $\exp(tA)$.

It remains to show why the semi-group T_t generated by A coincides with the function $\exp(tA)$ defined according to the spectral theory. The idea is to use Theorem 4.2 in Chapter II; we know that the Cauchy problem: $w'(t) = Aw(t)$, $w(0) = v_o \in D(A)$ is (uniformly) well-posed, in particular there is uniqueness of Cauchy problem. We shall see that if $h \in D(A)$, the vector function $z(t) = \exp(tA)h$ is a solution of $z'(t) = Az(t)$ such that $z(0) = h$ so that it will coincide with $T_t h$; using also density of $D(A)$ in H we find readily what we are looking for.

Hence, if we consider the ratio $\dfrac{z(t+\xi) - z(t)}{\xi}$, it is equal to $\int_{-\infty}^{0} \frac{1}{\xi} [e^{\lambda(t+\xi)} - e^{\lambda t}] dE_\lambda h$; when $\xi \to 0$, this has a limit which equals

$\int_{-\infty}^{0} \lambda e^{\lambda t} dE_\lambda h$ as can be proved using the dominated convergence theorem with respect to the measure induced by $(E_\lambda h, h)$ and the convergence of

$\int_{-\infty}^{0} |\lambda|^2 d(E_\lambda h, h)$ $(h \in D(A))$.

It can be also proved, using properties of these integrals, that

$z(t) \in D(A)$, $\forall t \geq 0$ (so that' $\int_{-\infty}^{0} |\mu|^2 d_\mu \| E_\mu \int_{-\infty}^{0} e^{\lambda t} dE_\lambda h \|^2 < \infty$) and that

$Az(t) = \int_{-\infty}^{0} \lambda e^{\lambda t} dE_\lambda h = z'(t)$, as requested.   Obviously also, for $t = 0$

holds $z(0) = \int_{-\infty}^{0} dE_\lambda h = h$.   Thus the Lemma is proved.

We shall now end the section, with statement and demonstration[*] of the

following

Theorem 2.4: *Let be $A \leq \Theta$ a self-adjoint operator, $D(A) \subseteq H \to H$, such*

that $Ax = \theta$ iff $x = \theta$.   Let us assume that $f_o \in H$ is not in the range

of A .   Then, if $v(t)$ is any solution of the equation $v' = Av + f_o$,

it follows that $\lim_{t \to \infty} \| v(t) \| = \infty$.

The proof will be by contradiction.   Let us assume there is a solution

$v_o(t)$ of the equation $v'_o = Av_o + f_o$, such that its norm is not convergent

to ∞ with t .   In this case we can find a sequence $(t_p)_1^\infty \to \infty$ such that:

$\sup_{p \in N} \| v_o(t_p) \| = N_o < \infty$.   We shall see that this is impossible.

Let us remark first that, using again the spectral theorem for A , it

is possible to find an increasing sequence $(E_n)_1^\infty$ of orthogonal projections

in H , such that

    (i) $\lim_{n \to \infty} E_n x = x$, $\forall x \in H$.

    (ii) A is reduced by the subspaces $E_n(H)$ and $\{x \in H, E_n x = \theta\}$.

    (iii) $E_n A E_n = A E_n$ considered as operator in $E_n(H)$ is bounded,

           symmetric, strictly negative (hence invertible too).

---

[*]due to Professor J. J. Schäffer (personal correspondence).

(We can take $E_n = E_{-\frac{1}{n}} - E_{-n}$ where $A = \int_{-\infty}^{0} \lambda dE_\lambda$) .

We see without difficulty that, denoting $v_n(t) = E_n v_o(t)$, $n = 1, 2, \ldots$ the functions $v_n(t)$ verify the equation

$$v_n'(t) = (AE_n)v_n(t) + E_n f_o , \quad n = 1, 2, \ldots \quad .$$

If we apply here Theorem 2.1 (AE_n being self-adjoint and strictly negative), we obtain that:

$$\lim_{t \to \infty} v_n(t) \quad \text{exists and} \quad - - (AE_n)^{-1} E_n f_o = x_n \quad \text{(by definition)}$$

In view of the relation $v_n(t) \in E_n(H)$ - which is a closed subspace of H - , it results that $x_n \in E_n(H)$ too.

Let us remark also: $\lim_{p \to \infty} v_n(t_p) = x_n$ (as $t_p \to \infty$) . On the other hand one has the estimate

$$\| E_n v_o(t_p) \| = \| v_n(t_p) \| \leq \| v_o(t_p) \| \leq N_o , \quad \forall p = 1, 2, \ldots , \quad n = 1, 2, \ldots$$

Using the equality $\| x_n \| = \lim_{p \to \infty} \| v_n(t_p) \|$, we find: $\| x_n \| \leq N_o,$ $\forall n \in N.$ By a well-known property of Hilbert spaces, we can extract a subsequence $(x_{n_q})_1^\infty$ which has weak limit $= x \in H$ when $q \to \infty$. Let us observe also that $E_n(H) \subset D(A)$, $n = 1, 2, \ldots$; thus $x_n \in D(A)$, $n = 1, 2, \ldots$ and also

$$A x_n = A E_n x_n = -(AE_n)(AE_n)^{-1} E_n f_o = -E_n f_o , \quad \forall n \in N \quad .$$

We have thus obtained: $(w) \lim_{q \to \infty} x_{n_q} = x$ and $\lim_{q \to \infty} A x_{n_q} = -f_o$. (By property (i) above). Take now any $h \in D(A) = D(A^*)$; we deduce $\lim_{q \to \infty} (A x_{n_q}, h) = - (f_o, h)$ or also

82

$$\lim_{q \to \infty} (x_{n_q}, Ah) = (x, Ah) = -(f_o, h), \forall h \in D(A) .$$

It results that $x \in D(A)$ and $Ax = -f_o$ so that f_o is in the range of A , a contradiction which proves the theorem.

6 Regularity of solutions

<u>INTRODUCTION</u>

In most of previous Chapters we have considered solutions of the equation $u' - Au = \theta$ or $u' - Au = f$ which were differentiable or continuously differentiable (see especiallly §4 of Chapter I where rudiments of "regularity type" results are given); in Chapter IV we have even considered "weak" solutions which are only square-integrable over an interval and need not possess any supplementary derivative. In this Chapter instead we shall explain a few results (without trying at all to be exhaustive on the matter), where, under additional(natural)assumptions, for example on the Cauchy data, it is possible to obtain additional regularity properties, like k-differentiability, ∞-differentiability or even analyticity of the solutions. Main references here are Hille [15], Gelfand [12] and E. Nelson [30].

§1.   <u>ON k-AND ∞-DIFFERENTIABILITY</u>

Let us reconsider here the Cauchy problem: $u'(t) = Au(t)$, $u(0) = u_o \in D(A)$ on the non negative half-line R^+ ; we shall assume that it is a uniformly well-posed problem; this will imply a representation formula: $u(t) = U(t)u_o$ $U(t)$ being a semi-group of class C_o (see Chapter II). If A is the infinitesimal generator of $U(t)$ (as happens when A is a linear *closed* operator in the underlying Banach space X) and if one takes an arbitrary element $x \in D(A)$, the vector-valued function $u(t) = U(t)x$ is continuously differentiable, $R^+ \to X$, verifies $u(0) = x$ and also the equality $u'(t) = Au(t) = U(t)(Ax)$, $\forall t \geq 0$.

In order to obtain differentiability of higher order we shall prove now the following (simple)

<u>Theorem 1.1</u>: *Let be* $x \in D(A^k)$, $k \geq 2$. *Then the function* $u(t) = U(t)x$ *is*
k-times continuously differentiable, $R^+ \to X$; *moreover,* $u(t)$ *belongs to*
$D(A^k)$, $\forall t \geq 0$, *and the equality*

$$u^{(k)}(t) = A^k u(t) = U(t)(A^k x) , \quad \forall t \geq 0$$

is satisfied.

<u>Proof</u>: Let us take as a starting point the already established equality for
$x \in D(A)$; $u'(t) = U(t)(Ax)$; therefore, if $x \in D(A^2)$, then $Ax \in D(A)$
and from the above remarks the function $u'(t)$ is of class $C^1(R^+;X)$ and
$u''(t) = U(t)(A^2 x)$.

In order to prove that $u(t) = U(t)x \in D(A^2)$ for $x \in D(A^2)$ we use what we
proved at the end of Chapter I: if $y \in D(A) \Rightarrow U(t)y \in D(A)$ and
$U(t)(Ay) = AU(t)y$. Actually, $x \in D(A^2) \Longleftrightarrow x \in D(A)$, and $Ax \in D(A)$.
From $x \in D(A) \Rightarrow U(t)x \in D(A)$, and $A(U(t)x) = U(t)(Ax)$. From $Ax \in D(A) \Rightarrow$
$U(t)(Ax) \in D(A)$ and $A(U(t)(Ax)) = U(t)(A^2 x)$. As $U(t)(Ax) = A(U(t)x)$,
we see that $A(U(t)x) \in D(A)$ and $A^2(U(t)x) = A(U(t)(Ax)) = U(t)(A^2 x)$.
Also, $u''(t) = U(t)(A^2 x) = A^2(U(t)x)$ from the above.

In general, for all k , we shall use induction; assume the result true
for k , and prove it for $k + 1$; take $x \in D(A^{k+1}) \Leftrightarrow x \in D(A^k)$ and
$A^k x \in D(A)$. Hence $U(t)x$ is k-times continuously differentiable;
$\dfrac{d^k}{dt^k}(U(t)x) = U(t)(A^k x)$ by the induction hypothesis. But $A^k x \in D(A) \Rightarrow$
$U(t)(A^k x) \in C^1(R^+;X)$, so that $\dfrac{d^{k+1}}{dt^{k+1}}(U(t)x) = U(t)(A^{k+1} x) \in C(R^+;X)$. To
prove: $U(t)x \in D(A^{k+1})$ we proceed as before: $x \in D(A^k) \Rightarrow U(t)x \in D(A^k)$
and $A^k(U(t)x) = U(t)(A^k x)$. As $A^k x \in D(A)$, we see that $U(t)(A^k x) \in D(A)$

and $A[U(t)(A^k x)] = U(t)(A^{k+1}x)$.   Hence $A^k(U(t)x) \in D(A)$ and $A^{k+1}(U(t)x)$

$= U(t)(A^{k+1}x)$ as requested.   Finally $u^{(k+1)}(t) = \frac{d}{dt} u^{(k)} = \frac{d}{dt} U(t)(A^k x) =$

$U(t)(A^{k+1}x)$ — because $A^k x \in D(A)$.   Hence $u^{(k+1)}(t) = A^{k+1}u(t) = U(t)(A^{k+1}x)$

thus ending the proof.

<u>Corollary</u>: *Let us assume that $x \in D(A^k)$, $\forall k = 1,2,3,\ldots$.   Then $u(t) =$*
$U(t)x \in C^\infty(R^+;X)$, $u(t) \in D(A^k)$, $\forall k = 1,2,3,\ldots$, and the equality
$u^{(n)}(t) = Au^{(n-1)}(t)$ is true $\forall n = 1,2,\ldots$, $\forall t \geq 0$.

In fact, $u^{(n)}(t) = A^n u(t) = A(A^{n-1}u(t)) = Au^{(n-1)}(t)$ from the Theorem.

An interesting result (see Hille [15]) is expressed in the following

<u>Theorem 1.2</u>: *If A is the infinitesimal generator of a C_o semi-group U_t ,*
there exists a dense set M in X , such that all solutions $u(t)$ of
$u' - Au = \theta$, $u(0) \in M$ are infinitely differentiable on R^+ .

<u>Proof</u>: Using the *Corollary* above it will suffice to show the existence of a
dense set M in X , such that $M \subseteq \overset{\infty}{\underset{k=1}{\cap}} D(A^k)$.   We shall use the following

<u>Lemma</u>: *Let be $K(t)$, $R^1 \to R^1$ a C^∞-function with compact support in the*
open interval $(0,\infty)$.   Then, $\forall x \in X$, the element $y = \int_o^\infty K(t)U(t)xdt$
belongs to $\overset{\infty}{\underset{k=1}{\cap}} D(A^k)$.

We shall prove that $y \in D(A^k)$, $\forall k = 1,2,\ldots$, firstly for $k = 1$,
then for any k by induction.   Let $h > 0$; we have: $\frac{1}{h}[U(h) - I] \int_o^\infty K(t)$
$U(t)xdt = \frac{1}{h} \int_o^\infty K(t)U(t+h)xdt - \frac{1}{h} \int_o^\infty K(t)U(t)xdt = \frac{1}{h} \int_h^\infty K(s-h)U(s)xds$

$- \frac{1}{h} \int_o^\infty K(s)U(s)xds = \int_o^\infty \frac{1}{h}[K(s-h) - K(s)]U(s)xds$.   (In view of: $K(s-h) = 0$

for $s \leq h$).   This tends to $- \int_o^\infty K'(s)U(s)xds$ as readily seen; hence:

$$\int_0^\infty K(t)U(t)x\,dt \; \epsilon \; D(A) \quad \text{and} \quad A\int_0^\infty K(t)U(t)x\,dt = -\int_0^\infty K'(t)U(t)x\,dt \;, \; \forall x \; \epsilon \; X.$$

Let us assume now (induction hypothesis), that $y \; \epsilon \; D(A^n)$ and that

$$A^n y = (-1)^n \int_0^\infty K^{(n)}(t)U(t)x\,dt \; . \quad \text{Consider again for} \quad h > 0 \;, \quad \text{the differential}$$

ratio $\frac{1}{h}[U(h) - I](-1)^n \int_0^\infty K^{(n)}(t)U(t)x\,dt$ and continue as above with

$(-1)^n K^{(n)}(t)$ in place of $K(t)$: but this is also a $C_0^\infty(0,\infty)$ function,

hence the above is true: $(-1)^n \int_0^\infty K^{(n)}(t)U(t)x\,dt \; \epsilon \; D(A)$ and

$$A(A^n y) = (-1)^{n+1} \int_0^\infty K^{(n+1)}(t)U(t)x\,dt \quad \text{as required.}$$

We have thus proved that the set M composed of all elements y of the

form $\int_0^\infty K(t)U(t)x\,dt$ where $K(t)$ is an arbitrary function in $C_0^\infty(0,\infty)$ and x

an arbitrary element in X , is included in the intersection $\bigcap_{k=1}^\infty D(A^k)$.

It remains to show that M is dense in X .

Let us consider a sequence of non-negative functions $K_n(t) \; \epsilon \; C_0^\infty(-\infty,\infty)$,

such that $K_n(t) = 0$ for $-\infty < t \leq \frac{1}{n}$ and for $t \geq \frac{2}{n}$, and also

$\int_{\frac{1}{n}}^{\frac{2}{n}} K_n(t)\,dt = 1$. We shall prove now that, given any element $x \; \epsilon \; X$, it

follows: $\lim\limits_{n\to\infty} y_n = \lim\limits_{n\to\infty} \int_0^\infty K_n(t)U(t)x\,dt = x$. In fact, we have

$$\left\| \int_0^\infty K_n(t)U(t)x\,dt - x \right\| = \left\| \int_0^\infty K_n(t)[U(t)x - U(0)x]\,dt \right\|$$

$$\leq \int_{\frac{1}{n}}^{\frac{2}{n}} K_n(t)\left\| U(t)x - U(0)x \right\|\,dt < \epsilon$$

for $n \geq N(\epsilon)$, as easily seen.

This ends the proof of Theorem 1.2.

§2. ANALYTIC SOLUTIONS

Let us consider functions $x(t)$, from a real open interval (a,b) into a
Banach space X ; we say that $x(t)$ is analytic if

(i) $x(t) \in C^{\infty}[(a,b);X]$ (indefinitely - strongly -

differentiable in (a,b))

and

(ii)   for any $t_o \in (a,b)$, $x(t)$ admits, in a subinterval

$|t-t_o| < h$ - where h may vary with t_o - a

representation $x(t) = \sum_{n=o}^{\infty} \frac{1}{n!} (t-t_o)^n x^{(n)}(t_o)$ in

the sense of strong convergence.

We shall also consider holomorphic functions $x(\zeta)$, from an open set Ω
of the complex plane into a Banach space X .   These are functions which
have a complex strong derivative $\forall \zeta \in \Omega$, precisely $\lim_{\lambda \to \zeta} \frac{1}{\lambda - \zeta}[x(\lambda) - x(\zeta)] =$
$x'(\zeta)$.   It can be proved (see [15]) that in any open disc:
$\{\zeta, |\zeta - \zeta_o| < R, \zeta_o \in \Omega\}$ which is completely contained in Ω ,
$x(\zeta) = \sum_{n=o}^{\infty} a_n(\zeta - \zeta_o)^n$ where $a_n = \frac{1}{n!} x^{(n)}(\zeta_o)$ - the existence of all

derivatives $x^{(k)}(\zeta_o) \ \forall \ \zeta_o \in \Omega$ being also a consequence of the existence of
the first derivative.

In particular, if $\Omega \subseteq C$ is an open set which contains the real open
interval $(a,b) \subseteq R$, and if $x(\zeta)$, $\Omega \to X$ is holomorphic in Ω , then,
its restriction to (a,b) is analytic as above defined.   A natural problem
for solutions $u(t)$ of a differential equation $u'(t) = Au(t)$ in an open
interval $(a,b) \subseteq R$, is precisely the problem of analyticity, for example

88

for a good choice of the initial data. A classical result in this respect
is the following (by I. Gelfand [12]).

Theorem 2.1: *Let be A a linear, closed, densely defined operator in the
Banach space X , such that the Cauchy problem: $u'(t) = Au(t)$, $u(0) \in D(A)$
is uniformly well-posed on the whole real line. Then, there exists a dense
subset of X , X_o , such that all solutions $u(t)$ with $u(\ ) \in X_o$ are
analytic on R .*

Proof: From previous Chapters we know the existence of a group G_t , $-\infty <
t < \infty \to L(X,X)$, which is strongly continuous $\forall t \in R$, has infinitesimal
generator $= A$, and permits representation of any solution $u(t)$ under the
form $u(t) = G_t u(o)$, $t \in R$; moreover, if $x \in D(A)$, $G_t x$ is a solution
on R .   Hence, our problem consists in finding a dense subset $X_o \subseteq X$ such
that $G_t x$ is analytic on $R, \forall x \in X_o$.   Let us associate to G_t two semi-
groups of class C_o :

$$G_1(t) , [o,\infty) \to L(X;X) , \quad G_1(t) = G(t) , \quad \forall t \geq 0$$
$$G_2(t) , [o,\infty) \to L(X;X) , \quad G_2(t) = G(-t) , \quad \forall t \geq 0 \quad .$$

If we use Theorem 2.1 in Chapter II, we find estimates

$$\| G_1(t) \| \leq M_1 e^{\omega_1 t} , \quad \| G_2(t) \| \leq M_2 e^{\omega_2 t} , \quad 0 \leq t < \infty \quad .$$

Accordingly:

$$\| G(s) \| \leq M_1 e^{\omega_1 s} , \quad 0 \leq s < \infty , \quad \| G(-s) \| \leq M_2 e^{\omega_2 s} , \quad 0 \leq s < \infty$$

or

$$\| G(s) \| \leq M_2 e^{-\omega_2 s} , \quad -\infty < s \leq 0 ,$$

or also

$$\| G(s) \| \leq M_2 e^{\omega_2 |s|} \quad , \quad -\infty < s \leq 0 \ .$$

Hence

$$\| G(s) \| \leq M e^{\omega |s|} \quad , \quad -\infty < s < \infty \ ,$$

where $M = \max(M_1, M_2)$, $\omega = \max(\omega_1, \omega_2)$.

Let us consider now, given $x \in X$ and $\nu > 0$, the improper integral

$$y_\nu = \frac{1}{\sqrt{4\pi\nu}} \int_{-\infty}^{\infty} e^{-\frac{s^2}{4\nu}} G(s)x \, ds \quad ;$$

this is a convergent integral as follows from the above estimate for $\| G(s) \|$. Moreover, if we use continuity of $G(s)x$, we deduce readily that: $\lim_{\nu \to 0} y_\nu = x$. Therefore, the set $X_0 = \{y_\nu \ , \ \nu > 0 \ , \ x \in X\}$ is dense in X , and the theorem will be proved once we have established the analyticity of $G(t)y_\nu \ , \forall \, y_\nu \in X_0$. In fact, we shall see that $G(t)y_\nu$ is the restriction to $z \in R$ of the (holomorphic) function

$$y_\nu(z) = \frac{1}{\sqrt{4\pi\nu}} \int_{-\infty}^{\infty} e^{-\frac{(s-z)^2}{4\nu}} G(s)x \, ds$$

in the complex plane. The convergence and holomorphy of this improper integral can be proved for example considering the infinite series:

$$\sum_{n=-\infty}^{\infty} \frac{1}{\sqrt{4\pi\nu}} \int_{n}^{n+1} e^{-\frac{(s-z)^2}{4\nu}} G(s)x \, ds \ .$$

Each of the integrals

90

$$g_{n,\nu}(z) = \frac{1}{\sqrt{4\pi\nu}} \int_n^{n+1} e^{-\frac{(s-z)^2}{4\nu}} G(s)x\,ds$$

is holomorphic for any $z \in C$. We shall now estimate $\| g_{n,\nu}(z) \|$. If $z = a + ib$, $s - z = s - a - ib$, $(s-z)^2 = (s-a)^2 - b^2 - 2ib(s-a)$, $\exp(-\frac{1}{4\nu}(s-z)^2) = \exp(\frac{1}{4\nu}(b^2 - (s-a)^2)) \cdot \exp(\frac{2i}{4\nu}b(s-a))$; hence

$$\| g_{n,\nu}(z) \| \leq \frac{1}{\sqrt{4\pi\nu}} \left(\int_n^{n+1} e^{\frac{b^2-(s-a)^2}{4\nu}} Me^{\omega|s|}ds \right) \| x \|$$

$$= \frac{M\|x\|}{\sqrt{4\pi\nu}} \int_n^{n+1} e^{\frac{b^2}{4\nu}} e^{\omega|s|} e^{-\frac{(s-a)^2}{4\nu}}\,ds .$$

Assume, say, that $n \in N$, $n > a$ and $|b| < L$; then

$$\| g_{n,\nu}(z) \| \leq \frac{M\|x\|}{\sqrt{4\pi\nu}} e^{\frac{L^2}{4\nu}} e^{\omega(n+1)} e^{-\frac{(n-a)^2}{4\nu}} = \frac{M\|x\|}{\sqrt{4\pi\nu}} e^{\frac{L^2}{4\nu}} e^{\omega(n+1)} e^{-\frac{n^2}{4\nu}}$$

$$e^{-\frac{a^2}{4\nu}} e^{\frac{2na}{4\nu}} \leq C e^{\omega n} e^{-\frac{n^2}{4\nu}} e^{\frac{2na}{4\nu}} .$$

Let us assume also that $|a| \leq A$; then the series $\sum_{n=1}^{\infty} e^{\omega n} e^{\frac{2na}{4\nu}} e^{-\frac{n^2}{4\nu}}$ is obviously uniformly convergent; similar reasonings hold for $n = -1, -2, \dots$ Consequently, we can estimate $\| g_{n,\nu}(z) \|$ in the rectangle $|b| < L$, $|a| < A$, by a convergent numerical series, uniformly for z in this rectangle; this proves holomorphy of $y_\nu(z)$ in the (arbitrary) rectangle $\{|a| < A, |b| < L\}$, hence in the complex plane too. To end the proof it is sufficient to show that

$$G(t)y_\nu = y_\nu(t) = \frac{1}{\sqrt{4\pi\nu}} \int_{-\infty}^{\infty} e^{-\frac{(s-t)^2}{4\nu}} G(s)x\,ds .$$

But, in fact, we see that

$$G(t)y_\nu = \frac{1}{\sqrt{4\pi\nu}} \int_{-\infty}^{\infty} e^{-\frac{s^2}{4\nu}} G(t+s)x\,ds$$

which equals $y_\nu(t)$ as is obtained after the substitution $\sigma = s + t$.

A natural problem (explicitly stated by E. Hille in [15]) is to ask whether the result expressed in Theorem 2.1 above can be extended to uniformly well-posed Cauchy problem - i.e. corresponding to C_o-semi-groups - on the positive half-line, (precisely, if given any C_o-semi-group T_t , $t \geq 0 \to L(X,X)$, it is possible to find a dense subset of X , X_o , such that $T_t x$ be analytic in (o,∞) for any $x \in X_o)$. A nice counterexample, solving the problem in the negative sense, was given by E. Nelson in [30].

Let us consider the Hilbert space $L^2(\mathrm{R}^+)$ consisting of complex-valued (measurable) functions $\varphi(s)$, $0 \leq s < \infty$ such that $\int_{\mathrm{R}^+} |\varphi(s)|^2 ds < \infty$. If $\varphi(\cdot) \in L^2(\mathrm{R}^+)$ and $t \geq 0$ we define an operator $S(t)$ through the formula: $(S(t)\varphi)(s) = \varphi(s-t)$ if $s \geq t$, $= 0$ if $0 \leq s < t$. This operator maps $L^2(\mathrm{R}^+)$ into itself; moreover:

$$\| S(t)\varphi \|^2_{L^2(\mathrm{R}^+)} = \int_t^{\infty} |\varphi(s-t)|^2 ds = \|\varphi\|^2_{L^2(\mathrm{R}^+)} \; ;$$

$S(t)$ is obviously a linear map on $L^2(\mathrm{R}^+)$ and $S(0)\varphi = \varphi$, $\forall \varphi \in L^2(\mathrm{R}^+)$; $S(t)$ is also norm-continuous $\forall t \geq 0$ and is an operator semi-group. The last property follows because: if $t_1, t_2 \in \mathrm{R}^+$, then $(S(t_1 + t_2)\varphi)(s) = \varphi(s - t_1 - t_2)$ for $s \geq t_1 + t_2$, $= 0$ for $0 \leq s < t_1 + t_2$. Furthermore, $(S(t_2)\varphi)(s) = \varphi(s - t_2)$, if $s \geq t_2$, $= 0$ if $0 \leq s < t_2$. Denote $(S(t_2)\varphi)(s)$ by $\psi_{t_2}(s)$. Therefore, $[S(t_1)S(t_2)\varphi](s) = \psi_{t_2}(s - t_1)$ if $s \geq t_1$, $= 0$ if $0 \leq s < t_1$. On the other hand, $\psi_{t_2}(s-t_1) = \varphi(s-t_1-t_2)$

92

if $s - t_1 \geq t_2$, $= 0$ if $0 \leq s-t_1 < t_2$. Hence, $(S(t_1)S(t_2)\varphi)(s) =$
$\varphi(s - t_1 - t_2)$ if $s \geq t_1 + t_2$, $= 0$ if $0 \leq s < t_1$, $= 0$ if $0 \leq s < t_1+t_2$,
so that equality with $(S(t_1 + t_2)\varphi)(s)$ is obvious. The former property
(norm-continuity) is proved also: for $t \geq 0$, $h > 0$, using the semi-group
property, we deduce relation

$$\| S(t+h)\varphi - S(t)\varphi \|_{L^2(\mathbb{R}^+)} = \| S(t)(S(h)\varphi-\varphi) \|_{L^2(\mathbb{R}^+)} = \| S(h)\varphi - \varphi \|_{L^2(\mathbb{R}^+)}.$$

Actually, $(S(h)\varphi)(s) = \varphi(s-h)$, $s \geq h$, $= 0$, $s < h$; hence $(S(h)\varphi - \varphi)(s)$
$= \varphi(s-h)-\varphi(s)$, $s \geq h$, $-\varphi(s)$, $0 \leq s < h$, and

$$\| S(h)\varphi-\varphi \|^2_{L^2(\mathbb{R}^+)} = \int_h^\infty |\varphi(s-h)-\varphi(s)|^2 ds + \int_0^h |\varphi(s)|^2 ds$$

$$= \int_0^\infty |\varphi(\xi) - \varphi(\xi+h)|^2 d\xi + \int_0^h |\varphi(\xi)|^2 d\xi \quad .$$

These integrals $\to 0$ as $h \downarrow 0$ by elementary properties of square-integrable
functions.

Finally, for $t > 0$ and $h < 0$ and small (so that $t + h > 0$), we can
write: $t + h = t' > 0$, $t = t' + |h|$, and therefore $S(t+h)\varphi - S(t)\varphi = S(t')\varphi$
$- S(t'+|h|)\varphi$, $\| S(t+h)\varphi - S(t)\varphi \|_{L^2(\mathbb{R}^+)} = \| S(t'+|h|)\varphi - S(t')\varphi \|_{L^2}$ which
$\to 0$ as $h \to 0$ by the above computations.

We have thus found that $S(t)$ is a C_0-semi-group on $L^2(\mathbb{R}^+)$. Let us
assume now that $\varphi(s)$ is a function in $L^2(\mathbb{R}^+)$ such that $S(t)\varphi$ is analytic
(as $L^2(\mathbb{R}^+)$ - valued function), in the open interval $(0,\infty)$. Remark also
that a Banach-valued function analytic in a real (open) interval is also
weakly analytic - as follows directly from the definition. In particular,
let us consider a function $\psi(s) \in L^2(\mathbb{R}^+)$ which vanishes for $s \geq s_o$; the

scalar product $(S(t)\varphi,\psi)_{L^2(\mathbb{R}^+)}$ equals $\int_t^\infty \varphi(s-t)\psi(s)ds = \int_0^\infty \varphi(\xi)\psi(t+\xi)d\xi$,

and by our assumption this integral, as a function of t, is analytic on

$(0,\infty)$.

Furthermore, $\psi(t+\xi) = 0$ when $t + \xi \geq s_o$, hence for $t \geq s_o$ too.

Hence, the analytic function $(S(t)\varphi,\psi)_{L^2(\mathbb{R}^+)}$ vanishes for $t \geq s_o$. By

well-known properties of (scalar) analytic functions on real intervals, we

deduce that $(S(t)\varphi,\psi)_{L^2(\mathbb{R}^+)} = 0$, $\forall t > 0$. Now, remark that the functions

$\psi \in L^2(\mathbb{R}^+)$ which are zero for $s \geq s_o$ (s_o being an arbitrary positive

number) form a dense subset of $L^2(\mathbb{R}^+)$. We obtain therefore that

$(S(t)\varphi,\psi)_{L^2(\mathbb{R}^+)} = 0$, $\forall t > 0$, $\forall \psi \in L^2(\mathbb{R}^+)$. This implies: $\| S(t)\varphi \|_{L^2(\mathbb{R}^+)} = 0$,

$\forall t > 0$. When $t \to 0$, we obtain $\|\varphi\|_{L^2(\mathbb{R}^+)} = 0$. Hence, for non-zero

initial data, the Cauchy problem for $u' = Au$ on $[0,\infty)$ (A being the

infinitesimal generator of the above considered C_o-semi-group $S(t)$) has

never analytic solutions on $(0,\infty)$.

Thus, in view of Theorem 2.1 and of this example, we deduce that the Cauchy

problems which are uniformly well-posed on the whole real-line behave quite

differently, as regards the problem of analyticity, than

Cauchy problems which are uniformly well-posed only on the positive half-line.

7 Asymptotic behavior (sequel)

<u>INTRODUCTION</u>

In the first section of this chapter we explain a very interesting result of
J. Goldstein [14] concerning bounded solutions of some second order nonhomogeneous equations in Hilbert spaces; it is a result which is similar to
"Esclangon's Lemma" in the theory of ordinary differential equations[1].   In
the next paragraph we shall explain some results concerning asymptotic
behavior of solutions of the equation $u'(t) = Au(t)$ with self-adjoint non-
positive unbounded operator A , following the fundamental work of P. D. Lax
[22], and we shall conclude with (another) asymptotic property for second-
order equations which is due to S. Rosencrans [35] and was kindly communicated
to us by the author some time ago.

1.   <u>BOUNDED SOLUTIONS OF THE SECOND-ORDER EQUATION</u>

We shall prove here the following

Theorem 1.1: *Let be* B *a self-adjoint operator with domain* $D(B)$ *in the
Hilbert space* H *; assume that* $u(t)$ *,* $t \in R \to D(B^2)$ *is twice continuously
differentiable in* H *, while* $u'(t) \in D(B)$ *, and* $Bu'(t)$ *is H-continuous*
$\forall t \in R$ *; let also* $f(t)$ *,* $t \in R \to H$ *be strongly continuous and equal to*
$u''(t) + B^2 u(t)$ *.   Then, the relations* $\sup_{t \in R} \| u(t) \| < \infty$ *,* $\sup_{t \in R} \| Bu(t) \| < \infty$ *,*

$\sup_{t \in R} \| f(t) \| < \infty$ *,* *imply* $\sup_{t \in R} \| u'(t) \| < \infty$ *too .*

---

[1] When writing this § we have used a preliminary version of [14], communicated
to us by Professor J. Goldstein.

$\underline{\text{Proof}}$: Remark first the equality: $u'(t) = \int_o^t [f(s) - (B^2 u)(s)]ds + u'(0)$.

Next, consider, $\forall \alpha > 0$, the operators $C_\alpha = iB - \alpha I$ $(i = \sqrt{-1})$.

Therefore, $\forall x \in D(B^2)$, holds $C_\alpha^2 x = -B^2 x - 2\alpha i B x + \alpha^2 x$.   Consider also

the exponential operator: $\exp(tC_\alpha) = e^{-\alpha t}\exp(itB)$, where $\exp(itB)$ is

defined for example, using Stone's theorem (see [34]).   Remark that

$\forall h \in D(B)$ holds $\frac{d}{dt} \exp(tC_\alpha)h = e^{-\alpha t}\exp(itB)(-\alpha h + iBh)$.

Let us consider also the vector-function: $\exp(tC_\alpha)g(t)$, where $g(t)$,

$t \in R \to D(C_\alpha)$ is H-continuously differentiable.   It is easy to prove that

$\frac{d}{dt} \exp(tC_\alpha)g(t)$ exists and $= \exp(tC_\alpha)g'(t) + \exp(tC_\alpha)C_\alpha g(t)$.

Let us remark now that $u'(t) - C_\alpha u(t) \in D(C_\alpha)$, since $u'(t) \in D(B) =$

$D(C_\alpha)$ and $u(t) \in D(B^2) = D(C_\alpha^2)$.   Let us prove that $(C_\alpha u)(t)$ is

continuously differentiable; we have

$$\frac{1}{\delta}\ [(C_\alpha u)(t+\delta) - (C_\alpha u)(t)] = C_\alpha \frac{1}{\delta}\ [u(t+\delta) - u(t)]$$

$$= C_\alpha \frac{1}{\delta} \int_t^{t+\delta} u'(\sigma)d\sigma = \frac{1}{\delta} \int_t^{t+\delta} C_\alpha u' d\sigma$$

because C_α is a closed operator, $u'(t) \in D(C_\alpha)$, $C_\alpha u'(\sigma)$ is H-continuous.

It follows that $C_\alpha u$ is differentiable and $(C_\alpha u)'(t) = C_\alpha u'(t)$.

Therefore, the function $u'(t) - (C_\alpha u)(t)$ is H-continuously differentiable

and $u' - C_\alpha u \in D(C_\alpha)$; moreover, the function $\exp(tC_\alpha)[u'(t) - (C_\alpha u)(t)]$

is differentiable and $\frac{d}{dt} \exp(tC_\alpha)\ [u'(t) - (C_\alpha u)(t)] = \exp(tC_\alpha)[u''(t) -$

$C_\alpha u'(t)] + \exp(tC_\alpha)[C_\alpha u' - C_\alpha^2 u] = \exp(tC_\alpha)(u'' - C_\alpha^2 u) = \exp(tC_\alpha)[u'' + B^2 u +$

$2\alpha i Bu - \alpha^2 u] = \exp(tC_\alpha)(f + 2\alpha i Bu - \alpha^2 u)$.   Let us take then

$-\infty < \tau < \sigma < +\infty$ and let us integrate between τ and σ .   We find the

equality

$$\exp(\sigma C_\alpha)[u'(\sigma) - (C_\alpha u)(\sigma)] - \exp(\tau C_\alpha)[u'(\tau) - (C_\alpha u)(\tau)]$$

$$= \int_\tau^\sigma \exp(t C_\alpha)[f(t) + 2\alpha i B u(t) - \alpha^2 u(t)]\,dt$$

and the integrand is estimated by $Ce^{-\alpha t}$ (using boundedness of f, Bu and u over the real line). Therefore, the improper integral: $\int_\tau^\infty \exp(t C_\alpha)[f(t) + 2\alpha i B u(t) - \alpha^2 u(t)]\,dt$ is absolutely convergent and this implies the existence of the (strong) limit: $\lim_{\sigma\to\infty} \exp(\sigma C_\alpha)[u'(\sigma) - (iB)u(\sigma) + \alpha u(\sigma)]$, $\forall \alpha > 0$. Again from boundedness of Bu and u and from the estimate $\|\exp(\sigma C_\alpha)\| = e^{-\alpha\sigma}$ we obtain that: $\lim_{\sigma\to\infty} \exp(\sigma C_\alpha)(Bu(\sigma)) = \lim_{\sigma\to\infty} \exp(\sigma C_\alpha)u(\sigma) = \theta$. Therefore, we conclude the existence of the (strong) limit: $\lim_{\sigma\to\infty} \exp(\sigma C_\alpha)u'(\sigma)$, $\alpha > 0$, and we can also write: $\exp(\sigma C_\alpha)u'(\sigma) = e^{-\alpha\sigma}$ $\exp(i\sigma B)u'(\sigma) = e^{-\alpha\sigma}g(\sigma)$, where $g(\sigma) = \exp(i\sigma B)u'(\sigma)$.

Let us put: $h(\alpha) = \lim_{\sigma\to\infty} e^{-\alpha\sigma}g(\sigma)$, $\forall \alpha > 0$. For any $0 < \varepsilon < \alpha$, we have $\alpha = \alpha - \varepsilon + \varepsilon$ and $e^{-\alpha\sigma}g(\sigma) = e^{-\varepsilon\sigma}e^{-(\alpha-\varepsilon)\sigma}g(\sigma)$; on the other hand $\lim_{\sigma\to\infty} e^{-(\alpha-\varepsilon)\sigma}g(\sigma) = h(\alpha-\varepsilon)$; therefore: $\lim_{\sigma\to\infty} e^{-\alpha\sigma}g(\sigma) = \lim_{\sigma\to\infty} (e^{-\varepsilon\sigma})$.

$h(\alpha-\varepsilon) = \theta$ and, accordingly, $\lim_{\sigma\to\infty} \exp(\sigma C_\alpha)u'(\sigma) = \theta$. We can infer about the equality $-\exp(\tau C_\alpha)[u'(\tau) - (C_\alpha u)(\tau)] = \int_\tau^\infty \exp(t C_\alpha)[f(t) + 2\alpha i(Bu)(t) - \alpha^2 u(t)]\,dt$ or also: $u'(\tau) = (C_\alpha u)(\tau) - \exp(-\tau C_\alpha)\int_\tau^\infty \exp(t C_\alpha)[f(t) + 2\alpha i(Bu)(t) - \alpha^2 u(t)]\,dt$. From this (final) representation we deduce boundedness of $u'(\tau)$ on R, as follows: $C_\alpha u = iBu - \alpha u$ is bounded on R by hypothesis; also $\|\exp(-\tau C_\alpha)\| = e^{\alpha\tau}$ and $\|\exp(t C_\alpha)[f(t) + 2\alpha i(Bu)(t) - \alpha^2 u(t)]\| \le Ce^{-\alpha t}$; hence: $\|u'(\tau)\| \le L + e^{\alpha\tau}\int_\tau^\infty Ce^{-\alpha t}\,dt = L + \dfrac{C}{\alpha}$ which proves the Theorem.

§2.   <u>BEHAVIOR AT ∞</u>

Let us consider again (as in Chapter V, §2), a self-adjoint operator A, in the Hilbert space H, such that $(Ah, h) \leq 0$, $\forall h \in D(A)$. Then, as seen previously, A is the infinitesimal generator of the semi-group $S(t) = \exp(tA) = \int_{-\infty}^{0} e^{\lambda t} dE_\lambda$, where $\{E_\lambda\}_{-\infty}^{\infty}$ is the spectral resolution to A, and any solution of the equation: $u'(t) - Au(t) = \theta$, on $0 \leq t < \infty$ is represented by the formula: $u(t) = S(t)u(0)$.

Let us define now, for any fixed $u_o \in D(A)$, the numerical set $E_{u_o} = \{\lambda \in R, E_\lambda u_o = u_o\}$. In our case $(A \leq \theta)$, we have $E_\lambda = I$, $\forall \lambda \geq 0$, hence $0 \in E_{u_o}$; define $\lambda_o(u_o) = \inf E_{u_o}$; hence $\lambda_o(u_o) \leq 0$. We shall now prove the following

<u>Lemma</u>:   If $\lambda_o < \lambda^* \leq 0$, then $\lambda^* \in E_{u_o}$.

In fact, because of the definition of λ_o as exact (greatest) lower bound, we can find $\bar\lambda \in E_{u_o}$, $\lambda_o \leq \bar\lambda < \lambda^*$, so that $E_{\bar\lambda} u_o = u_o$ and accordingly, $(E_{\bar\lambda} u_o, u_o) = \|E_{\bar\lambda} u_o\|^2 = \|u_o\|^2$. On the other hand, the scalar-valued function of λ: $\phi_{u_o}(\lambda) = \|E_\lambda u_o\|^2$ is non-decreasing (as follows from the definition of the spectral resolution); therefore: $\|E_{\bar\lambda} u_o\|^2 \leq \|E_{\lambda^*} u_o\|^2 \leq \|u_o\|^2$, and $\|E_{\lambda^*} u_o\|^2 = \|u_o\|^2$; now, $u_o = E_{\lambda^*} u_o + (I - E_{\lambda^*})u_o \Rightarrow \|u_o\|^2 = \|E_{\lambda^*} u_o\|^2 + \|(I - E_{\lambda^*})u_o\|^2 = \|u_o\|^2 + \|(I - E_{\lambda^*})u_o\|^2 \Rightarrow \|(I - E_{\lambda^*})u_o\| = 0 \Rightarrow E_{\lambda^*} u_o = u_o$; accordingly, $\lambda^* \in E_{u_o}$, thus ending the proof of Lemma.

<u>Corollary</u>: $E_{\lambda_o} u_o = u_o$ too, because $\{E_\lambda\}$ is right-continuous, so that

$$E_{\lambda_o} u_o = \lim_{\lambda \downarrow \lambda_o} E_\lambda u_o = u_o \ .$$

Remark at this stage that if $u(t) = \int_{-\infty}^{0} e^{\lambda t} dE_\lambda u_o$ then $u(t) = \int_{-\infty}^{\lambda_o} e^{\lambda t} dE_\lambda u_o$,

as $\int_{\lambda_o}^{0} e^{\lambda t} dE_\lambda u_o = \theta$ obviously; we are able now to prove the following

Theorem 2.1: *Given the initial data $u_o \in D(A)$, the solution $u(t)$ of*

$u'(t) - Au(t) = \theta$, $t \geq 0$, $u(0) = u_o$, *satisfies the following relations:*

$$u(t) = O(e^{(\lambda_o + \varepsilon)t}), \quad \forall \varepsilon > 0 \quad (when \ t \to +\infty)$$

$$u(t) \neq O(e^{(\lambda_o - \varepsilon)t}), \quad \forall \varepsilon > 0 \quad (when \ t \to +\infty)$$

Proof: Let us take any $\varepsilon > 0$; then

$$\| u(t)e^{-(\lambda_o + \varepsilon)t} \|^2 = \int_{-\infty}^{\lambda_o} e^{2\lambda t} e^{-2\lambda_o t} e^{-2\varepsilon t} d\| E_\lambda u_o \|^2$$

$$= e^{-2\varepsilon t} \int_{-\infty}^{\lambda_o} e^{2(\lambda - \lambda_o)t} d\| E_\lambda u_o \|^2 \leq e^{-2\varepsilon t} \| u_o \|^2$$

which $\to 0$ as $t \to +\infty$.

In order to prove the second statement of the theorem, let us write:

$$\| u(t)e^{-(\lambda_o - \varepsilon)t} \|^2 = \int_{-\infty}^{0} e^{2(\lambda - \lambda_o)t} e^{2\varepsilon t} d\| E_\lambda u_o \|^2$$

$$\geq e^{2\varepsilon t} \int_{\lambda_o - \delta}^{\lambda_o} e^{2(\lambda - \lambda_o)t} d\| E_\lambda u_o \|^2,$$

where $\delta > 0$.

Also, $\lambda > \lambda_o - \delta$, $t > 0$, $\Rightarrow (\lambda - \lambda_o)t > -\delta t$, so that

$$\|u(t)e^{-(\lambda_o-\varepsilon)t}\|^2 \geq e^{2\varepsilon t}\, e^{-2\delta t} \int_{\lambda_o-\delta}^{\lambda_o} d\|E_\lambda u_o\|^2 \ .$$

Now, the scalar function $\|E_\lambda u_o\|^2$ is not constant on $[\lambda_o-\delta\,,\,\lambda_o]$ (otherwise, $\|E_\lambda u_o\|^2 = \|u_o\|^2$ for $\lambda_o-\delta \leq \lambda \leq \lambda_o$; this would imply, as previously, that $E_\lambda u_o = u_o$ for these λ , contradicting the definition of λ_o). Therefore, $\int_{\lambda_o-\delta}^{\lambda_o} d\|E_\lambda u_o\|^2$ is > 0 , and $\|u(t)e^{-(\lambda_o-\varepsilon)t}\|^2 \geq$

$C_o e^{2(\varepsilon-\delta)t} = C_o e^{\varepsilon t}$ (taking $\delta = \frac{\varepsilon}{2}$) which proves the second statement of the theorem[1].

This Theorem has the following interesting

Corollary: *Let be* $u(t)$ *,* $0 \leq t < \infty$ *a solution of the equation* $u'(t) = Au(t)$ *where* A *is self-adjoint* $\leq \theta$ *, and let us assume that* $\lim\limits_{t\to\infty} e^{bt}\|u(t)\| = 0$ *for all real numbers* b *. Then* $u(t) = \theta$ *,* $\forall t \geq 0$ *.*

In fact, our hypothesis means: $u(t) = (e^{bt})$, $t \to +\infty$, $\forall b \in R$. If $\lambda_o(u(0))$ given by the Theorem is $> -\infty$, we obtain an obvious contradiction. Hence, $E_\lambda u(0) = u(0)$, $\forall \lambda \leq 0$. Also, $\lim\limits_{\lambda\to-\infty} E_\lambda u(0) = \theta$ so that $u(0) = \theta$, and $u(t) = \theta,\ \forall t \geq 0$ as required.

We shall finish this Chapter with a result concerning an asymptotic property for differential equations of second order. Precisely, let us state the

Theorem 2.2: *Let be* A *;* $D(A) \subseteq H \to H$ *, a symmetric operator, such that* $(Ax,x) \geq a^2\|x\|^2$ *,* $\forall x \in D(A)$ *,* $a > 0$ *. Let also* $u(t)$ *,* $0 \leq t < \infty \to D(A)$ *be a twice continuously differentiable function, such that* $u''(t) - Au(t) = \theta$

---

[1] And even more: $\|u(t)e^{-(\lambda_o-\varepsilon)t}\| \to \infty$ when $t \to +\infty$, $\forall \varepsilon > 0$.

100

on $[0,\infty)$ and $Re(u(0),u'(0)) > 0$.   Then, there exists a constant $C > 0$
such that $\|u(t)\| > Ce^{at}$ for large t .

Proof: Let $\varphi(t) = (u(t),u(t)) = \|u(t)\|^2$.   Then $\varphi'(t) = 2\,Re(u(t),$
$u'(t)) \leq 2\|u(t)\|\,\|u'(t)\| = 2\|u'(t)\|\sqrt{\varphi(t)}$.

Then it is also

$$\varphi''(t) = \frac{d}{dt}\,[(u',u) + (u,u')] = (u'',u) + (u,u'') + 2\|u'(t)\|^2$$

$$= 2\|u'(t)\|^2 + 2(Au(t),\,u(t)) \geq 2\|u'(t)\|^2 + 2a^2\|u(t)\|^2$$

$$= 2\|u'(t)\|^2 + 2a^2\varphi(t) \geq 4a\|u'(t)\|\sqrt{\varphi(t)}$$

(using the elementary inequality: $2\alpha\beta \leq \alpha^2 + \beta^2$,   where $\alpha = \sqrt{2}\|u'(t)\|$,
$\beta = a\sqrt{2}\sqrt{\varphi(t)}$) .

Therefore we find the inequality

$$\varphi''(t) \geq 4a\|u'(t)\| \cdot \sqrt{\varphi(t)} \geq 2a\,\varphi'(t) .$$

Integrating from 0 to $t > 0$, we obtain

$$\varphi'(t) - \varphi'(0) \geq 2a\varphi(t) - 2a\varphi(0) ,$$

hence

$$\varphi'(t) - 2a\varphi(t) \geq \varphi'(0) - 2a\varphi(0) .$$

Let us write then

$$\varphi'(t) - 2a\varphi(t) - \varphi'(0) + 2a\varphi(0) = \varepsilon(t) \geq 0$$

hence

$$\varphi'(t) - 2a\,\varphi(t) = \varphi'(0) - 2a\,\varphi(0) + \varepsilon(t) \ ,$$

where $\varepsilon \geq 0$.

If one solves this differential equation, one finds

$$\varphi(t) = e^{2at}\varphi(0) + \int_0^t e^{2a(t-\sigma)}(\varphi'(0) - 2a\,\varphi(0) + \varepsilon)\,d\sigma$$

$$= e^{2at}\varphi(0) - \frac{1}{2a}(\varphi'(0) - 2a\,\varphi(0)\ )(1 - e^{2at}) + \int_0^t e^{2a(t-\sigma)}\varepsilon(\sigma)\,d\sigma$$

$$= e^{2at}\varphi(0) + \frac{1}{2a}e^{2at}(\varphi'(0) - 2a\,\varphi(0)) - \frac{1}{2a}(\varphi'(0) - 2a\,\varphi(0))$$

$$+ \int_0^t e^{2a(t-\sigma)}\varepsilon(\sigma)\,d\sigma \ .$$

Accordingly

$$\varphi(t) \geq e^{2at}\varphi(0) + \frac{1}{2a}e^{2at}(\varphi'(0) - 2a\,\varphi(0)) - \frac{1}{2a}(\varphi'(0) - 2a\,\varphi(0))$$

$$= \frac{1}{2a}e^{2at}\varphi'(0) - \frac{1}{2a}\varphi'(0) + \varphi(0) \ ,$$

that is

$$\varphi(t) \geq \frac{1}{2a}\varphi'(0)[e^{2at} - 1] + \varphi(0) \ .$$

Also $\varphi'(0) = 2\,Re(u(0)\,,\,u'(0)) > 0$, and we deduce therefore:
$\varphi(t) \geq \beta\,e^{2at} - \beta + \varphi(0)$, $\beta > 0$, hence $\|u(t)\|^2\,e^{-2at} \geq \beta - \beta e^{-2at} +$
$\varphi(0)e^{-2at} \geq \frac{\beta}{2}$ for $t \geq t_0$. Hence

$$\|u(t)\|\,e^{-at} \geq \frac{\sqrt{\beta}}{\sqrt{2}} \quad \text{for} \quad t \geq t_0 \ ,$$

which proves the Theorem.

8 Weak solutions on the real line

<u>INTRODUCTION</u>

In this Chapter we consider again weak solutions of first-order abstract differential equations as were defined in Chapter IV.   The results explained here are closely connected with our previous work [41,[42].   There we insisted mainly on the so called "global existence theorems" (i.e. existence of solutions on the whole real line) and a main tool in the proof was a certain density result.   Here, we shall concentrate ourselves on such a density (or approximation) theorem, which in more concrete cases was considered by C. Runge (see [8]) and B. Malgrange, [25].   We shall follow quite closely our paper [47] where the approximation property is studied in a rather abstract setting (see also [28] for a related result).

§1.   <u>PRELIMINARIES</u>

Again, as in Chapter IV, we shall consider a Hilbert space H , equipped with a scalar product $(,)$ and an induced norm $\| h \|^2 = (h,h)$, $\forall h \in H$.
Next, let A be a linear closed operator with dense domain $D(A) \subseteq H$; then let A^* be its adjoint, so that $(Ah,k) = (h,A^*k)$ is valid $\forall h \in D(A)$ and $\forall k \in D(A^*)$, the domain $D(A^*)$ consisting of those $k \in H$ such that (Ah,k) can be represented as (h,k^*) with $k^* \in H$.

Given an open interval $(a,b) \subseteq R$ the class of test functions $K_{A^*}(a,b)$ (vector-valued) is defined as in Chapter IV.

We take now an arbitrary positive number T and we denote by V_T the set of weak solutions in $(-T,T)$ of the (homogeneous) differential equation:

$$u'(t) = Au(t) \ , \quad \text{so that} \quad \int_{-T}^{T} [(u(t),\varphi'(t)) + (u(t),A^*\varphi(t))]dt = 0 \ ,$$

$\forall \varphi \ \varepsilon \ K_{A^*}(-T,T) \ ,$ where $u(t) \ \varepsilon \ B^2(-T,T;H) \ .$ We also denote by V_∞ the class of functions $u(t) \ \varepsilon \ B^2_{loc}(R;H)$ verifying the relation $\int_R (u(t) , \varphi'(t) + A^*\varphi(t))dt = 0 \ , \forall \varphi \ \varepsilon \ K_{A^*}(R) \ .$ We shall now give the following

<u>Definition 1</u>: The abstract differential operator $\dfrac{d}{dt} - A$ has the approximation property if, for any pair of positive numbers $T_1 < T_2$, the set V_∞ is dense in V_{T_2} in the norm of $B^2(-T_1,T_1;H)$.

(This means that, given any $u \ \varepsilon \ V_{T_2}$ and $\varepsilon > 0$, we can find an element $u_\varepsilon \ \varepsilon \ V_\infty$, such that $\int_{-T_1}^{T_1} \| u(t)-u_\varepsilon(t) \|^2 dt < \varepsilon^2$).

Now, consider (another) linear closed operator with dense domain in H , U and give the following

<u>Definition 2</u>: The abstract differential operator $\dfrac{d}{dt} - U$ has the support property if the following is true: given any couple of functions $u(t), f(t)$ belonging to $B^2_{loc}(R;H)$ such that: $u' - Uu = f$ on $-\infty < t < \infty$ in weak sense, u has compact support in R , f has support contained in the finite interval $[a,b]$, it results that the support of u is also contained in $[a,b]$.

The next statement defines "property (P)".

<u>Definition 3</u>: We say that property (P) is verified if, given any finite interval $(a,b) \subset R$, one may find a constant $C_{a,b}$ in such a way that the inequality

$$\| \psi \|_{B^2(a,b;H)} \leq C_{a,b} \| \psi' + A^*\psi \|_{B^2(a,b;H)} \ , \quad \forall \psi \ \varepsilon \ K_{A^*}(a,b)$$

is valid.

104

In the following section we shall dedicate ourselves to the proof of the following

Theorem 1.1: *Let us assume that the operator $\frac{d}{dt} + A^*$ has the support property and that property (P) is also verified. Then the operator $\frac{d}{dt} - A$ has the approximation property.*

§2.   APPROXIMATION PROPERTIES

In order to prove the above stated result, a principal tool will be the following

Main Lemma: *Let us assume the Hypothesis of Theorem 1.1 and let us consider three positive numbers $0 < T_1 < T_2 < T_3$. Then, the set V_{T_3} is dense in V_{T_2} in the norm of $B^2(-T_1, T_1; H)$. (This means, naturally, that $\forall u \in V_{T_2}$ and $\forall \varepsilon > 0$, there exists $u_\varepsilon \in V_{T_3}$, such that $\int_{-T_1}^{T_1} \| u(t) - u_\varepsilon(t) \|^2 \, dt < \varepsilon^2$).*

In order to prove this result we shall use the following criterion for density:

Proposition 1: *Let be H a Hilbert space, and $A \subset B \subset H$ two linear subspaces. Let us assume that $(h,a) = 0$, $\forall a \in A$ implies $(h,b) = 0$, $\forall b \in B$, where h is any element in H . Then A is dense in B .*

Proof: "A is dense in B" means that any element in B is limit of a sequence of elements $(a_n)_1^\infty \subset A$; so that $B \subset \bar{A} = $ closure of A . If A *is not dense* in B , A is not dense in the closure of B , $\bar{B} \subset H$, and this in turn implies that $\bar{A}$ is a *strict* closed linear subspace of $\bar{B}$. Using elementary Hilbert space theory we find an element $h \in \bar{B}$, $h \neq \theta$, and $h \perp \bar{A}$. Thus, $(h,a) = 0$, $\forall a \in A$ and (h,b) *is not* $= 0$, $\forall b \in B$ (this would imply $(h,b) = 0$, $\forall b \in \bar{B}$, therefore $(h,h) = 0$, absurd). *This proves the Proposition.*

When we apply this result to our situation we take H to be $B^2(-T_1, T_1; H)$ while A is the restriction to the interval $(-T_1, T_1)$ of the set V_{T_3} and B is the restriction to $(-T_1, T_1)$ of the set V_{T_2}; moreover, V_{T_3} and V_{T_2} are linear closed subspaces of $B^2(-T_3, T_3; H)$ and respectively $B^2(-T_2, T_2; H)$ (as easily seen from Chapter IV); as a consequence, A and B are linear (non necessarily closed) subspaces of $B^2(-T_1, T_1; H)$. Therefore, in view of the above it will suffice, to demonstrate the *Main Lemma*, proving that the relation: $\int_{-T_1}^{T_1} (h(t), v(t)) dt = 0$, $\forall v \in V_{T_3}$ implies the relation

$$\int_{-T_1}^{T_1} (h(t), w(t)) dt = 0, \quad \forall w \in V_{T_2}$$

where $h(t)$ is an arbitrary element in $B^2(-T_1, T_1; H)$. Let us define now the extended function $\tilde{h}(t)$, which $= h(t)$ on $(-T_1, T_1)$ and $= \theta$ if $|t| \geq T_1$.

Let us define also the set $M \subseteq B^2(-T_3, T_3; H)$ as the image of $K_{A^*}(-T_3, T_3)$ through the operator $\frac{d}{dt} + A^*$, i.e. $M = \{\varphi' + A^*\varphi, \varphi \in K_{A^*}(-T_3, T_3)\}$ (as in Chapter IV). We have

<u>Proposition 2</u>: *The function $\tilde{h}(t)$ belongs to the closure of M in $B^2(-T_3, T_3; H)$.*

<u>Proof</u>: Again, we shall use a simple criterion, valid in general Hilbert spaces H; precisely, if A is a linear subspace of H, an element $x \in H$ belongs to the closure of A iff the relation $(f, a) = 0$,

$\forall a \in A \Rightarrow (f,x) = 0$[1]   (the reader can check for himself this useful result).

Therefore, taking $H = B^2(-T_3,T_3;H)$ and $A = M$, let us assume that

$$\int_{-T_3}^{T_3} (f(t),\varphi'(t) + A^*\varphi(t))dt = 0 \quad \text{for} \quad f(t) \in B^2(-T_3,T_3;H) \ . \quad (\forall \varphi \in K_{A^*}(-T_3,T_3))$$

This means however that $f(t) \in V_{T_3}$; then

$$\int_{-T_3}^{T_3} (f(t),\tilde{h}(t))dt = \int_{-T_1}^{T_1} (f(t),h(t))dt = 0$$

by assumption. We infer that $\tilde{h}(t) \in \text{closure}(M)$, as required.

The result expressed in Proposition 2 means the existence of a sequence of functions $\varphi_n \in K_{A^*}(-T_3,T_3)$ such that $\int_{-T_3}^{T_3} \| \tilde{h}(t) - (\varphi_n' + A^*\varphi_n) \|^2 \, dt \to 0$

as $n \to \infty$. Hence, the sequence $(\varphi_n' + A^*\varphi_n)_1^\infty$ is a Cauchy sequence in

$B^2(-T_3,T_3;H)$ and if we apply property (P) we deduce that the sequence

$(\varphi_n)_1^\infty$ is also a Cauchy sequence in $B^2(-T_3,T_3;H)$; let be $\phi(t) = \lim_{n\to\infty} \varphi_n(t)$

in $B^2(-T_3,T_3;H)$. Define also the extended function: $\tilde{\phi}(t) = \phi(t)$,

$-T_3 < t < T_3$, $= \theta$, $|t| \geq T_3$, and let us prove now the following:

<u>Proposition 3</u>: *The support of $\tilde{\phi}$ is contained in the interval* $[-T_1,T_1]$.

In order to establish this fact, we show that $\frac{d}{dt}\tilde{\phi} + A^*\tilde{\phi} = \tilde{h}$ in weak sense

over R, and then apply the support property for $\frac{d}{dt} + A^*$. Remember that

the adjoint of A^* is A itself (see [37]) and consider the expression:

$$\int_{-\infty}^{\infty} (\tilde{\phi}(t), \xi'(t) - (A\xi)(t))dt \ , \quad \forall \xi \in K_A(R) \ .$$

---

[1]Where f is any element in H.

We have:

$$\int_R (\tilde{\phi}, \xi' - A\xi)\,dt = \int_{-T_3}^{T_3} (\phi(t), \xi'(t) - (A\xi)(t))\,dt = \lim_{n\to\infty} \int_{-T_3}^{T_3} (\varphi_n(t), \xi'(t) - (A\xi)(t))\,dt$$

$$= \text{(as easily seen, using integration by parts)} = -\lim_{n\to\infty} \int_{-T_3}^{T_3} (\varphi'_n(t) + (A^*\varphi_n)$$

$$(t), \xi(t))\,dt = -\int_{-T_3}^{T_3} (\tilde{h}(t), \xi(t))\,dt = -\int_R (\tilde{h}(t), \xi(t))\,dt, \quad \forall \xi \in K_A(R) \,, \quad \text{q.e.d.}$$

At this stage we are nearer to the proof of Main Lemma. Remember that, in view of Proposition 1, we *assume* that $\displaystyle\int_{-T_1}^{T_1} (h(t), v(t))\,dt = 0$, $\forall v \in V_{T_3}$ and *try to prove* that

$$\int_{-T_1}^{T_1} (h(t), w(t))\,dt = 0 \,, \quad \forall w \in V_{T_2} \,,$$

where $h(t) \in B^2(-T_1, T_1; H)$. We extend h as $\tilde{h} = \theta$ outside $[-T_1, T_1]$, and therefore we *must now prove* that:

$$\int_{-T_2}^{T_2} (\tilde{h}(t), w(t))\,dt = 0 \,, \quad \forall w \in V_{T_2} \,.$$

We shall use here the method of mollifers (adapted from scalar-case, as exposed in [36], [11]) to vector-valued case, in a convenient, natural way.

Let $\alpha(t)$ be a $C^1(R)$ (scalar-valued) function, with support in $|t| \leq 1$, such that $\alpha(t) \geq 0$, $\int_R \alpha(t)\,dt = 1$. Let $\alpha_\varepsilon(t) = \dfrac{1}{\varepsilon} \alpha(\dfrac{t}{\varepsilon})$. Thus,

$\alpha_\varepsilon(t) = 0$ for $|t| \geq \varepsilon$ and $\int_R \alpha_\varepsilon(t)\,dt = \dfrac{1}{\varepsilon} \int_R \alpha(\dfrac{t}{\varepsilon})\,dt = 1$, $\forall \varepsilon > 0$. Given an interval (a, b) and a function $z(t) \in B^2_{loc}(a, b; H)$ we define the convolution: $(z*\alpha_\varepsilon)(t) = \int_{t-\varepsilon}^{t+\varepsilon} z(\tau)\alpha_\varepsilon(t-\tau)d\tau$ which is (obviously) a $C^1(H)$

function defined for $a+\varepsilon < t < b-\varepsilon$; its strong derivative equals

$\int_{t-\varepsilon}^{t+\varepsilon} z(\tau)\alpha_\varepsilon'(t-\tau)d\tau$. Next, we take a sequence $\alpha_n(t) = n\alpha(nt)$ and we see

that $\int_R \alpha_n(t)dt = 1$. Consider the convolutions $(\tilde{h}*\alpha_n)(t)$ which are

$C^1(R;H)$ and our next step will be to demonstrate that $\int_{-T_2}^{T_2} (\tilde{h}*\alpha_n, w)dt = 0$,

$\forall n = 1,2,\ldots$, $w \in V_{T_2}$.

Consider also the convolution $\tilde{\phi}*\alpha_n$, likewise belonging to $C^1(R;H)$.
We *shall prove below* that, because $(\frac{d}{dt} + A*)\tilde{\phi} = \tilde{h}$ in weak sense over R ,
it results: $\tilde{\phi}*\alpha_n \in D(A*)$, $\forall t \in R$ and $(\tilde{\phi}*\alpha_n)' + A*(\tilde{\phi}*\alpha_n) = \tilde{h}*\alpha_n$ holds in
the usual sense, $\forall t \in R$.   (See [41],[42]).

Once we assume this property, we can replace

$$\int_{-T_2}^{T_2} (\tilde{h}*\alpha_n, w)dt \quad \text{by} \quad \int_{-T_2}^{T_2} ((\tilde{\phi}*\alpha_n)' + A*(\tilde{\phi}*\alpha_n), w)dt .$$

Furthermore, it is easy to see that $\tilde{\phi}*\alpha_n \in K_{A*}(-T_2, T_2)$ for n
sufficiently large $(\tilde{\phi} = \theta$ outside $[-T_1, T_1]$ implies that $\tilde{\phi}*\alpha_n$ whose
support is contained obviously in $[-T_1 - \frac{1}{n} , T_1 + \frac{1}{n}]$, has compact support
in $(-T_2, T_2)$ for $n \geq n_o)$.

Consequently, using appartnence of w to V_{T_2} we find that

$$\int_{-T_2}^{T_2} (\tilde{h}*\alpha_n, w)dt = \int_{-T_2}^{T_2} ((\tilde{\phi}*\alpha_n)' + A*(\tilde{\phi}*\alpha_n), w)dt = 0$$

for $n \geq n_o$.

We shall also prove below that $\tilde{h}*\alpha_n \to \tilde{h}$ in the space $B^2(-T_2, T_2 ; H)$.
Once we know this, we obtain:

$$\lim_{n\to\infty} \int_{-T_2}^{T_2} (\tilde{h}*\alpha_n, w)dt = \int_{-T_2}^{T_2} (\tilde{h}, w)dt = 0 \quad \forall w \; \varepsilon \; V_{T_2}$$

so that the Main Lemma is thus demonstrated.

Let us give now proofs for two results concerning mollifiers which were used above:

Firstly, let us write the equation: $\dfrac{d\tilde{\phi}}{dt} + A*\tilde{\phi} = \tilde{h}$ in weak sense over R . This means

$$\int_R (\tilde{\phi}(\tau), \xi'(\tau) - (A\xi)(\tau))d\tau = - \int_R (\tilde{h}(\tau), \xi(\tau))d\tau , \quad \forall \xi \; \varepsilon \; K_A(R) .$$

Let us take functions $\xi_n(\tau)$ of special form: $\xi_n(\tau) = \alpha_n(t - \tau)k$, where $k \; \varepsilon \; D(A)$ (we see that $\xi_n(\tau) \; \varepsilon \; K_A(R) , \forall n = 1, 2, \ldots$) , and t is taken in R. We remark that $\xi_n(\tau) = 0$ for $|\tau - t| > \dfrac{1}{n}$, and we obtain the relation

$$\int_R (\tilde{\phi}(\tau), -\alpha_n'(t-\tau)k - \alpha_n(t-\tau)Ak)d\tau = -\int_R (\tilde{h}(\tau), \alpha_n(t-\tau)k)d\tau$$

or also

$$\int_R (\tilde{\phi}(\tau)\alpha_n'(t-\tau), k)d\tau + \int_R (\tilde{\phi}(\tau)\alpha_n(t-\tau), Ak)d\tau = \int_R (\alpha_n(t-\tau)\tilde{h}(\tau), k)d\tau$$

and because of commutativity of integration with scalar product, we get

$$(\int_R \alpha_n'(t-\tau)\tilde{\phi}(\tau)d\tau, k) = (\int_R \alpha_n(t-\tau)\tilde{\phi}(\tau)d\tau , Ak) + (\int_R \alpha_n(t-\tau)\tilde{h}(\tau)d\tau, k) ,$$

$$\forall k \; \varepsilon \; D(A) ,$$

a relation that can be written as:

$$((\tilde{\phi}*\alpha_n)',k) + ((\tilde{\phi}*\alpha_n),Ak) = (\tilde{h}*\alpha_n,k) \, , \quad \forall k \in D(A) \ .$$

For any fixed $t \in R$ this equality implies: $\tilde{\phi}*\alpha_n \in D(A^*)$ and

$A^*(\tilde{\phi}*\alpha_n) = \tilde{h}*\alpha_n - (\tilde{\phi}*\alpha_n)' \,$, as required.

Next, we shall prove that $\tilde{h}*\alpha_n \to \tilde{h}$ in $B^2(-T_2,T_2;H)$. We use the Fourier transform which is possible because $\tilde{h}$ and $\tilde{h}*\alpha_n$ belong to $B^2(R;H)$ and we prove that $F(\tilde{h}*\alpha_n) \to F(\tilde{h})$ in $B^2(R;H)$. Actually $\tilde{h} \in B^1(R;H)$ and $F(\tilde{h}*\alpha_n) = F(\tilde{h}).F(\alpha_n)$ as easily seen.

Therefore:

$$\int_R \left| F(\tilde{h}*\alpha_n) - F(\tilde{h}) \right|^2 d\tau = \int_R \left| F(\tilde{h})(\sigma) \right|^2 \left| F(\alpha_n)(\sigma)-1 \right|^2 d\sigma \ .$$

From definition of α_n we see that $(F(\alpha_n))(\tau) = F(\alpha)(\frac{\tau}{n})$. Thus $F(\alpha_n)(\sigma) = F(\alpha)(\frac{\sigma}{n}) \to (F(\alpha))(0) = 1$ as $n \to \infty$, $\forall \sigma \in R$ and we can apply the dominated convergence theorem in order to get the result[1]. Let us give now the

Proof of Theorem 1.1: We take two positive numbers $T_1 < T_2$, a function $u(t) \in V_{T_2}$ and an $\varepsilon > 0$. We must find $u_\varepsilon \in V_\infty$ such that

$$(\int_{-T_1}^{T_1} \| u(t) - u_\varepsilon(t) \|^2 dt)^{\frac{1}{2}} < \varepsilon \ .$$

For this purpose, we shall consider an increasing sequence $T_3 < T_4 < \dots$ where $T_2 < T_3$ and $\lim_{n \to \infty} T_n = \infty$. We shall apply successively the Main Lemma to triplets $(T_1,T_2,T_3),(T_2,T_3,T_4),\dots$ etc. First we find a function $u_1(t) \in V_{T_3}$ in such a way that

---

[1] Here $F(f) = \int_R e^{-i\tau t} f(t)dt$ for $f \in B^1(R;H)$.

$\|u-u_1\|_{B^2(-T_1, T_1;H)} < \varepsilon/2$. Then, a function $u_2(t) \in V_{T_4}$ exists, such

that $\|u_1-u_2\|_{B^2(-T_2, T_2;H)} < \varepsilon/2^2$; continuing this way we find, $\forall n \in N$,

a function $u_n(t) \in V_{T_{n+2}}$, with the property that $\|u_{n-1}-u_n\|_{B^2(-T_n, T_n;H)} < \varepsilon/2^n$.

We see that $u_n(t) \in B^2(-T_{n+2}, T_{n+2};H)$, $\forall n \in N$. We can extend all u_n as $= \theta$ outside $(-T_{n+2}, T_{n+2})$: thus $u_n \in B^2(R;H)$, $\forall n = 1,2,\ldots$

We shall prove that the sequence $(u_n)_1^\infty$ is a Cauchy sequence in $B_{loc}^2(R;H)$. Choose hence a finite, arbitrary interval $[a,b]$; then, for sufficiently large n , it is $-T_n < a < b < T_n$. We must prove that $(u_k)_1^\infty$ is a Cauchy sequence in $B^2([a,b];H)$. For any $p \in N$ and for the above found n, we will have

$$\|u_n-u_{n+p}\|_{B^2(a,b;H)} \leq \|u_n-u_{n+1}\|_{B^2(-T_n, T_n;H)} + \|u_{n+1}-u_{n+2}\|_{B^2(-T_n, T_n;H)}$$

$$+ \ldots + \|u_{n+p-1}-u_{n+p}\|_{B^2(-T_n, T_n;H)}$$

which is obviously estimated by $\dfrac{\varepsilon}{2^{n+1}} + \dfrac{\varepsilon}{2^{n+2}} + \ldots \dfrac{\varepsilon}{2^{n+p}} \leq \dfrac{\varepsilon}{2^n}$; this can be

done $\leq$ any ε' if $n \geq n(\varepsilon')$, $\forall p = 1,2,\ldots$, thus proving that $(u_k)_1^\infty$ is a Cauchy sequence in $B^2(a,b;H)$ (note that $\varepsilon > 0$ is now fixed from the beginning of the proof).

We shall take now a sequence of intervals: $[-N,N]$, $N = 1,2,\ldots$; using completeness of $B^2(-N,N;H)$, $\forall N = 1,2,\ldots$, we can find functions $u^N(t) \in B^2(-N,N;H)$, such that $\|u_n(t) - u^N(t)\|_{B^2(-N,N;H)} \to 0$ as $n \to \infty$,

$\forall N = 1, 2\ldots\ldots$ It is obvious that $u^N(t) = u^M(t)$ almost everywhere on $[-M,M]$ in case $M < N$ so that we can define a function $u_\varepsilon(t)$ which equals

112

$u^N(t)$ in $[-N,N]$, $N = 1,2,\ldots$. It belongs to $B^2_{loc}(R;H)$ as readily seen, and it is a function in V_∞ because of the following: any function $\varphi(t) \in K_{A*}(R)$ will have support contained in some interval $(-N,N)$. Therefore

$$\int_R (u_\varepsilon(t),\varphi'(t) + (A^*\varphi)(t))dt = \int_{-N}^N (u^N(t),\varphi'(t) + (A^*\varphi)(t))dt$$

$$= \lim_{n\to\infty} \int_{-N}^N (u_n(t),\varphi'(t) + (A^*\varphi)(t))dt = 0$$

because for n big enough, $-T_{n+2} < -N < N < T_{n+2}$ and $u_n \in V_{T_{n+2}}$ while $\varphi(t) \in K_{A*}(-T_{n+2},T_{n+2})$.

Finally, $\|u-u_\varepsilon\|_{B^2(-T_1,T_1;H)}$ equals $\lim_{n\to\infty} \|u-u_n\|_{B^2(-T_1,T_1;H)}$. On the other hand, we can write

$$\|u-u_n\|_{B^2(-T_1,T_1;H)} \leq \|u-u_1\|_{B^2(-T_1,T_1;H)} + \|u_1-u_2\|_{B^2(-T_1,T_1;H)} + \ldots$$

$$\|u_{n-1}-u_n\|_{B^2(-T_1,T_1;H)} \leq \frac{\varepsilon}{2} + \frac{\varepsilon}{2^2} + \ldots \frac{\varepsilon}{2^n} \leq \varepsilon, \quad \forall n \in N .$$

Therefore $\|u-u_\varepsilon\|_{B^2(-T_1,T_1;H)} \leq \varepsilon$ as requested.

9 Elementary solutions

<u>INTRODUCTION</u>

Consideration of elementary (or fundamental) solutions of abstract differential operators $\frac{1}{i}\frac{d}{dt} - A$ (regardless of Cauchy problem) was proposed by Agmon-Nirenberg ([2]- pg. 122) and was subsequently studied by the author (in [43]); regularity properties and extensions to locally convex spaces were published in [27] and [4]; further extensions can be found in the recent paper [19].

In this Chapter we shall present, along with definitions in the framework of (elementary) distribution theory, a simple result concerning existence of fundamental solutions, under a polynomial growth condition for the resolvent $R(\lambda;A)$.

§1.   <u>DEFINITION OF ELEMENTARY SOLUTIONS</u>

Let us consider a Banach space X ; let A be a linear closed operator with dense domain $D_A \subset X$. Define $\mathcal{D}(R) = C_o^\infty(R)$ to be the linear space of infinitely differentiable scalar-valued functions $\varphi(t)$ with compact support in R . A sequence $(\varphi_p)_1^\infty \subset \mathcal{D}(R)$ is convergent to 0 if the supports supp φ_p are all contained in a fixed compact subset K of R , and if $\varphi_p^{(k)}(t) \to 0$ uniformly on K , $k = 0,1,2,\ldots$.

Next, define $\mathcal{D}'(R;X)$ (the space of X-valued distributions on R) consisting of all linear maps T from $\mathcal{D}(R)$ into X , such that
$$\lim_{p\to\infty} \| T(\varphi_p) \|_X = 0 , \quad \forall \text{ sequence } (\varphi_p)_1^\infty \text{ convergent to } 0 \text{ in } \mathcal{D}(R) .$$

Examples: (i) if $b(t)$, $R \to X$ is a locally integrable-Bochner function, define $T_b(\varphi) = \int_R b(t)\varphi^{(p)}_{(t)} dt$, $\forall \varphi \in \mathcal{D}(R)$; (ii) $(\delta \otimes I)(\varphi) = \varphi(0)I$,

$\forall \varphi \in \mathcal{D}(R)$, where I is the identity map of X , defines a linear map from

$\mathcal{D}(R)$ into $L(X;X)$ which is continuous in the above defined sense.

If we define derivatives of X-valued distributions by formula

$(D^k T)(\varphi) = (-1)^k T(\varphi^{(k)})$, $\forall \varphi \in \mathcal{D}(R)$, we see that any $T \in \mathcal{D}'(R;X)$ is

infinitely differentiable, thus $D^k T \in \mathcal{D}'(R;X)$, $\forall k \in N$.

Let us consider now the subset $D_A \subset X$ endowed with the graph topology:

$\| x \|_{D_A} = \| x \|_X + \| Ax \|_X$, $\forall x \in D_A$: it becomes thus a complete Banach

space, because of closedness of A . We shall also use the space $L(X;D_A)$

of all linear continuous maps: $X \to D_A$ so that

$$\| B \|_{L(X;D_A)} = \sup_{\| x \|_X \leq 1} \| Bx \|_{D_A} = \sup_{\| x \|_X \leq 1} \{ \| Bx \|_X + \| ABx \|_X \} \quad .$$

Let us remark now that A is a linear *continuous* map of D_A into X ; we

can consider then distributions $T \in \mathcal{D}'(R;L(X;D_A))$, and then define AT as

distribution in $\mathcal{D}'(R;L(X;X))$ by the formula

$$(AT)(\varphi) = A(T(\varphi)) , \quad \forall \varphi \in \mathcal{D}(R)$$

(AT is obviously well defined: $T(\varphi) \in L(X;D_A)$, $\forall \varphi \in \mathcal{D}(R)$; but

$A \in L(D_A;X)$; hence $A(T(\varphi)) \in L(X;X)$; it depends linearly on φ as easily

seen; also, if $\varphi_n \to 0$ in $\mathcal{D}(R)$, $\Rightarrow T(\varphi_n) \to \theta$ in $L(X;D_A)$, hence

$(AT)(\varphi_n) \to \theta$ in $L(X;X)$.

The few concepts introduced above suffice for a natural definition of

elementary solutions: we say that the distribution $E \in \mathcal{D}'(R;L(X;D_A))$ is an

elementary solution for the operator $\frac{1}{i}\frac{d}{dt} - A = L$ (for formal convenience,

in this Chapter we consider the operator $\frac{1}{i}\frac{d}{dt} - A$ rather than the previously

studied $\frac{d}{dt} - A$) if $LE = \delta \otimes I$, i.e. $\frac{1}{i}\frac{dE}{dt} - AE = \delta \otimes I$, i.e., $\forall \varphi \in \mathcal{D}(R)$,

$$- \frac{1}{i} E(\varphi') - AE(\varphi) = \varphi(o)I \quad \text{in} \quad L(X;X) \; .$$

One more remark is needed to justify this definition: $E \in \mathcal{D}'(R;L(X;D_A)) \Rightarrow \frac{dE}{dt} \in \mathcal{D}'(R;L(X;D_A)) \subseteq \mathcal{D}'(R;L(X;X)) \; .$

§2. EXISTENCE OF ELEMENTARY SOLUTIONS

As pointed out in the introduction, we shall indicate here a simple sufficient condition on the resolvent operator $(\lambda-A)^{-1}$ which ensures existence of a fundamental solution for the operator $\frac{1}{i} \frac{d}{dt} - A$; this condition is less restrictive than in [43], [27], following a suggestion in [19]; stronger assumptions on $R(\lambda;A)$ would imply (as indicated in [43]) further regularity properties of the fundamental solution thus constructed, but this study is beyond our present purpose.

Definition: Let be Λ the union: $(-\infty < \lambda \leq -N) \cup (N \leq \lambda < \infty)$ and Γ a regular simple arc in the complex plane, joining the points $(-N,0)$ and $(N,0)$ [1] .

We shall prove below the following

Theorem 2.1: *Let us assume that the resolvent operator* $(\lambda-A)^{-1}$ *belongs to* $L(X;X)$ *for* $\lambda \in \Lambda \cup \Gamma$. *Furthermore, an estimate* $\| (\lambda-A)^{-1} \|_{L(X;X)} \leq C(1+|\lambda|)^k$ *for* $\lambda \in \Lambda$, *where* k *is a positive integer, is verified. Then, the operator* $\frac{1}{i} \frac{d}{dt} - A$ *has a fundamental solution.*

Proof: It is known (see for example [37]) that the set $\rho(A)$ of all complex numbers λ such that $(\lambda-A)^{-1} \in L(X;X)$ is an open subset of the complex plane

---

[1] Thus $\Gamma = \{\lambda(t), \alpha \leq t \leq \beta \to C , \lambda(t) \in C^1[\alpha,\beta], \lambda(\alpha) = (-N,0), \lambda(\beta) = (N,0)\}$ $\lambda(t_1) \neq \lambda(t_2)$ if $t_1 \neq t_2$.

116

and that the function: $\lambda \in \rho(A) \to R(\lambda;A) \in L(X;X)$ is analytic - i.e. the strong complex derivative $R'(\lambda;A)$ exists $\forall \lambda \in \rho(A)$. As a corollary we derive the existence of $R(\lambda;A)$ in an open neighborhood of $\Lambda \cup \Gamma$ and its continuity there.

On the other hand: if $\varphi(t)$ is any function in $\mathcal{D}(R)$, its (inverse) Fourier transform: $\breve{\varphi}(\lambda) = (\sqrt{2\pi})^{-1} \int_R e^{i\lambda t}\varphi(t)dt$ extends to an entire analytic function of λ ; moreover, for real λ , a sequence of estimates:
$|\breve{\varphi}(\lambda)| \leq C_p(1+|\lambda|^2)^{-p}$, $\lambda \in R$, $p = 0,1,2,\dots$, is verified, as easily seen; finally, the Fourier inversion formula holds, i.e. $\varphi(t) = (\sqrt{2\pi})^{-1} \int_R e^{-i\lambda t}\breve{\varphi}(\lambda)d\lambda$ in particular, $\varphi(0) = (\sqrt{2\pi})^{-1} \int_R \breve{\varphi}(\lambda)d\lambda$.

We shall consider now, $\forall \varphi \in \mathcal{D}(R)$, the (improper) integral

$\dfrac{1}{\sqrt{2\pi}} \int_{\Lambda\cup\Gamma} \breve{\varphi}(\lambda)R(\lambda;A)d\lambda$.   (Its precise definition is, obviously:

$\dfrac{1}{\sqrt{2\pi}} [\lim_{R\to\infty} \int_{-R}^{-N} \breve{\varphi}(\lambda)R(\lambda;A)d\lambda + \lim_{R\to\infty} \int_{N}^{R} \breve{\varphi}(\lambda)R(\lambda;A)d\lambda + \int_{\alpha}^{\beta} \breve{\varphi}(\lambda(t))R(\lambda(t);A)\lambda'(t)dt])$.

The integrand is a continuous function; moreover, the estimate
$\|\breve{\varphi}(\lambda)R(\lambda;A)\| \leq C_p(1+|\lambda|^2)^{-p}\cdot C(1+|\lambda|)^k$, used for p sufficiently large, proves absolute convergence of the above integrals.)

The resolvent $R(\lambda;A)$ can be considered as a continuous function from $\Lambda\cup\Gamma$ to $L(X;X)$ but also to $L(X;D_A)$.   This last fact follows because:

$R(\lambda;A)x \in D_A$, $\forall x \in X$ and $\| R(\lambda;A)x \|_X + \| AR(\lambda;A)x \|_X = \| R(\lambda;A)x \|_X +$

$\| (A-\lambda)R(\lambda;A)x + \lambda R(\lambda;A)x \| \leq C(\lambda)\| x \| + \| x \| + |\lambda|C(\lambda)\| x \| = C_1(\lambda)\| x \|$,

and a few more similar estimates.

Thus, we can derive that the element $\dfrac{1}{\sqrt{2\pi}} \int_{\Lambda\cup\Gamma} \breve{\varphi}(\lambda)R(\lambda;A)d\lambda$ belongs to

$L(X;D_A)$, $\forall \varphi \in \mathcal{D}(R)$.   Therefore, we can define a mapping E , from $\mathcal{D}(R)$

to $L(X;D_A)$, which is given by the relation $\langle E,\varphi\rangle = \frac{1}{\sqrt{2\pi}} \int\limits_{\Lambda\cup\Gamma} \check{\varphi}(\lambda)R(\lambda;A)d\lambda$.

This is a linear map, because of linearity of the inverse Fourier transform $\varphi \to \check{\varphi}$ and of the integral.

Next we shall prove that $E \in \mathcal{D}'(R;L(X;D_A))$ and that $\frac{1}{i}\frac{dE}{dt} - AE = \delta \otimes I$.

For the first part we need to show that: $(\varphi_j)_1^\infty \subset \mathcal{D}(R)$, $\lim\limits_{j\to\infty}\varphi_j = 0$ in $\mathcal{D}$, implies $\langle E,\varphi_j\rangle \to 0$ in $L(X;D_A)$.

Remark that, as well known (see for example [5],[18]), a corollary of the convergence of φ_j to 0 in $\mathcal{D}(R)$ is the convergence of φ_j to 0 in $S(R)$ — space of C^∞ - rapidly decreasing functions - and hence of $\check{\varphi}_j$ to 0 in the same space.   Thus, if $C_p^j = \sup\limits_{\lambda\in R}(1+|\lambda|^2)^p|\check{\varphi}_j(\lambda)|$, then $\lim\limits_{j\to\infty}C_p^j = 0$, $\forall p = 0,1,2,\ldots$

On the other hand, we shall establish now an estimate of $\|\langle E,\varphi\rangle\|$ in $L(X;D_A)$, $\forall \varphi \in \mathcal{D}(R)$.

Consider any $x \in X$; we have $\langle E,\varphi\rangle x = \frac{1}{\sqrt{2\pi}} \int\limits_{\Lambda\cup\Gamma} \check{\varphi}(\lambda)R(\lambda;A)x d\lambda$.   Here the integrand belongs to D_A : in fact $A(\check{\varphi}(\lambda)R(\lambda;A)x) = \check{\varphi}(\lambda)(A-\lambda+\lambda)R(\lambda;A)x = -\check{\varphi}(\lambda)x + \lambda\check{\varphi}(\lambda)R(\lambda;A)x$, which is again an integrable function on $\Lambda\cup\Gamma$, as easily seen.   We obtain the equality:

$$A\langle E,\varphi\rangle x = \frac{1}{\sqrt{2\pi}} \int\limits_{\Lambda\cup\Gamma} \lambda\check{\varphi}(\lambda)R(\lambda;A)x d\lambda - \frac{1}{\sqrt{2\pi}} \int\limits_{\Lambda\cup\Gamma} \check{\varphi}(\lambda)x d\lambda .$$

Then we have estimates for $\|\langle E,\varphi\rangle x\|_X + \|A\langle E,\varphi\rangle x\|_X = \|\langle E,\varphi\rangle x\|_{D_A}$ as follows: $\|\langle E,\varphi\rangle x\|_X \leq \frac{1}{\sqrt{2\pi}} \int\limits_{\Lambda} |\check{\varphi}(\lambda)|\,\|R(\lambda;A)x\|\,d\lambda + \frac{1}{\sqrt{2\pi}} \int\limits_{\Gamma} |\check{\varphi}(\lambda)|\,\|R(\lambda;A)x\|\,|d\lambda| = I_1 + I_2$; then $I_1 \leq \frac{1}{\sqrt{2\pi}}\|x\| \int\limits_{|\lambda|>N} C_p(1+|\lambda|^2)^{-p}C(1+|\lambda|)^k d\lambda = C_1\cdot C_p\|x\|$

(choosing p large enough) $(C_p = C_p(\varphi) = \sup\limits_{\lambda\in R}(1+|\lambda|^2)^p|\check{\varphi}(\lambda)|)$.   For I_2 we write: $I_2 = \frac{1}{\sqrt{2\pi}} \int\limits_\alpha^\beta |\check{\varphi}(\lambda(t))|\,\|R(\lambda(t);A)x\|\,|\lambda'(t)|dt$.

Now: from the definition $\check{\varphi}(z) = \frac{1}{\sqrt{2\pi}} \int_R e^{izt}\varphi(t)dt$ we obtain; if $[-A,A]$ is an interval containing the support of φ, and $z = x + iy$, then

$$\check{\varphi}(z) = \frac{1}{\sqrt{2\pi}} \int_{-A}^{A} e^{ixt} e^{-yt}\varphi(t)dt \;,\quad |\check{\varphi}(z)| \leq \sup_{|t|\leq A} |\varphi(t)| e^{|y|A}\left(\frac{2A}{\sqrt{2\pi}}\right)\;.$$

Thus $|\check{\varphi}(\lambda(\xi))| \leq \sup\limits_{|t|\leq A} |\varphi(t)| e^{|\nu(\xi)|A} \cdot \frac{2A}{\sqrt{2\pi}}$, $\forall\, \xi \in [\alpha,\beta]$; where

$\lambda(\xi) = \mu(\xi) + i\nu(\xi)$. But $\nu(\xi)$ is continuous, hence bounded on $[\alpha,\beta]$. We find an absolute constant $K_A > 0$, such that $|\check{\varphi}(\lambda(\xi))| \leq K_A \sup\limits_{|t|\leq A} |\varphi(t)|$ where $[-A,A]$ is an interval containing the support of φ.

Next, $R(\lambda;A)$ is a continuous, hence a bounded function on Γ - which is a compact in C. It follows: $\|R(\lambda(t);A)x\| \leq \sup\limits_{\alpha\leq t\leq \beta} \|R(\lambda(t);A)\| \, \|x\| = C_1\|x\|$. We get finally, an estimate $I_2 \leq C_{1,A}\|x\| \sup\limits_{|t|\leq A} |\varphi(t)|$, where $C_{1,A}$ is a positive constant. Consequently, $\|<E,\varphi>x\|_X \leq \|x\| \, (C_1 \sup\limits_{\lambda\in R}(1+|\lambda|^2)^p|\check{\varphi}(\lambda)|$ $+ C_{1,A} \sup\limits_{|t|\leq A} |\varphi(t)|)$ where p is a fixed positive integer; this estimate shows that, for a sequence $(\varphi_j) \subset D$, such that $\varphi_j \to 0$ in D, we have

$\lim\limits_{j\to\infty} <E,\varphi_j>x = \theta$ in X, uniformly for $\|x\|_X \leq 1$, that is

$\lim\limits_{j\to\infty} \|<E,\varphi_j>\|_{L(X,X)} = 0$.

Consider now an estimate for $\|A<E,\varphi>x\|$. The integral

$\frac{1}{\sqrt{2\pi}} \int_{\Lambda\cup\Gamma} \lambda\check{\varphi}(\lambda)R(\lambda;A)xd\lambda$ decomposes again in $\frac{1}{\sqrt{2\pi}} \int_{\Lambda} \lambda\check{\varphi}(\lambda)R(\lambda;A)xd\lambda + \frac{1}{\sqrt{2\pi}} \int_{\Gamma}$

$\lambda\check{\varphi}(\lambda)R(\lambda;A)xd\lambda = T_3 + T_4$. Now for p large $\|T_3\| \leq \|x\| \int_{|t|\geq N}$

$C_p(1+|t^2|)^{-p+\frac{1}{2}} \cdot C(1+|t|)^k dt = C'_p\|x\|$, and $C'_p \to 0$ for $\varphi \to 0$ in $D(R)$

and also $\|T_4\| \leq \frac{1}{\sqrt{2\pi}} \int_{\alpha}^{\beta} |\lambda(\xi)| |\check{\varphi}(\lambda(\xi))| \, \|R(\lambda(\xi);A)x\| \, |\lambda'(\xi)|d\xi \leq (C'_1 \sup\limits_{|t|\leq A} |\check{\varphi}(t)|) \|x\|$

as before, and again, it follows that

$$T_4 = T_4(\varphi) \to 0 \quad \text{as} \quad \varphi \to 0 \quad \text{in} \quad \mathcal{D}(R) .$$

The last integral to be considered is: $\frac{1}{\sqrt{2\pi}} \int_{\Lambda \cup \Gamma} \check{\varphi}(\lambda)x\,d\lambda = T_5$. We estimate it by: $\|T_5\| \leq \|x\| \, [\, \int_{\Lambda} |\check{\varphi}(\lambda)|\,d\lambda + \int_{\Gamma} |\check{\varphi}(\lambda)|\,|d\lambda| \,]$, and then we conclude as above.

These estimates are all uniform for $\|x\|_X \leq 1$; thus we get $\|<E,\varphi_j>\|_{L(X;D_A)} \to 0$ if $\varphi_j \to 0$ in $\mathcal{D}(R)$, therefore $E \in \mathcal{D}'(R;L(X;D_A))$ as required.

The final part of the proof deals with computation of LE , that is $\frac{1}{i}\frac{dE}{dt} - AE$.

First of all, we have

$$<\frac{1}{i}\frac{dE}{dt},\varphi> = -\frac{1}{i}<E,\varphi'> = -\frac{1}{\sqrt{2\pi}}\frac{1}{i} \int_{\Lambda \cup \Gamma} (\check{\varphi}')(\lambda)R(\lambda;A)\,d\lambda .$$

Now, we remark that

$$(\check{\varphi}')(\lambda) = \frac{1}{\sqrt{2\pi}} \int_{R} e^{i\lambda t}\varphi'(t)\,dt = -\frac{1}{\sqrt{2\pi}} \int_{R} (i\lambda)e^{i\lambda t}\varphi(t)\,dt = -i\lambda\check{\varphi}(\lambda) .$$

Therefore

$$<\frac{1}{i}\frac{dE}{dt},\varphi> = \frac{1}{\sqrt{2\pi}} \int_{\Lambda \cup \Gamma} \lambda\check{\varphi}(\lambda)R(\lambda;A)\,d\lambda .$$

Next, we compute AE which is given by the formula $(AE)(\varphi) = A(E(\varphi)) = A\frac{1}{\sqrt{2\pi}} \int_{\Lambda \cup \Gamma} \check{\varphi}(\lambda)R(\lambda;A)\,d\lambda$; remember that $R(\lambda;A)$ is a continuous function of λ with range in $L(X;D_A)$ and that $A \in L(D_A;X)$. Then $AR(\lambda,A)$ is a continuous function of λ with range in $L(X;X)$ and $A\frac{1}{\sqrt{2\pi}} \int_{\Lambda \cup \Gamma} \check{\varphi}(\lambda)R(\lambda;A)\,d\lambda = \frac{1}{\sqrt{2\pi}} \int_{\Lambda \cup \Gamma} \check{\varphi}(\lambda)AR(\lambda;A)\,d\lambda = \frac{1}{\sqrt{2\pi}} \int_{\Lambda \cup \Gamma} \check{\varphi}(\lambda)(A-\lambda+\lambda)R(\lambda;A)\,d\lambda = \frac{1}{\sqrt{2\pi}} \int_{\Lambda \cup \Gamma} \check{\varphi}(\lambda)[\lambda R(\lambda;A) - I]\,d\lambda .$

We derive:

$$\langle \frac{1}{i} \frac{dE}{dt} , \varphi \rangle - \langle AE, \varphi \rangle = (\frac{1}{\sqrt{2\pi}} \int_{\Lambda \cup \Gamma} \breve{\varphi}(\lambda)d\lambda)I \ .$$

Finally, remark that $\dfrac{1}{\sqrt{2\pi}} \int_{\Gamma} \breve{\varphi}(\lambda)d\lambda = \dfrac{1}{\sqrt{2\pi}} \int_{-N}^{N} \breve{\varphi}(\lambda)d\lambda$ because $\breve{\varphi}(\lambda)$ is analytic $\forall\ \lambda\ \varepsilon\ C$; then

$$\frac{1}{\sqrt{2\pi}} \int_{\Lambda \cup \Gamma} \breve{\varphi}(\lambda)d\lambda = \frac{1}{\sqrt{2\lambda}} \int_{-\infty}^{\infty} \breve{\varphi}(\lambda)d\lambda = \varphi(0)$$

as remarked earlier.   This proves our Theorem.

10 Almost-periodic solutions

<u>INTRODUCTION</u>

In this Chapter we shall explain a certain result concerning almost periodic solutions of the nonhomogeneous differential equation: $u'(t) = Au(t) + f(t)$, A being again a linear unbounded operator in a Banach space while $f(t)$ is a (vector-valued) almost-periodic function.

Precisely, we shall prove existence and uniqueness of the almost-periodic solution $u(t)$, when A is the infinitesimal generator of a C_o-semigroup with exponential decay as $t \to \infty$ (as in [39], [32]).

1. <u>EXISTENCE AND UNIQUENESS OF AN ALMOST-PERIODIC SOLUTION</u>

First we shall give Bohr-Bochner's definition of almost-periodicity for Banach space valued functions.

Let $f(t)$, $t \in R \to X$ - a Banach space - be strongly continuous and satisfy also the following property: *for any* $\varepsilon > 0$, *there exists* $L(\varepsilon) > 0$ *such that in any interval* $[a, a+L] \subset R$ *one may find a number* τ_ε, *in such a way that* $\sup\limits_{t \in R} \| f(t+\tau_\varepsilon) - f(t) \|_X < \varepsilon$.

Then we say that $f(t)$ is X-almost-periodic. Almost-periodic functions possess many interesting properties; we shall need only a few of them and for proofs we shall refer to Amerio-Prouse [3], or to the pioneering memoir of Bochner [6] who was the first to give a detailed - and exhaustive - study of almost-periodic functions with values in a Banach space.

Let now $S(t)$, $t \in R^+ \to L(X,X)$ be a C_o-operator semi-group, verifying an estimate: $\| S(t) \| \leq Me^{\beta t}$, $\forall t \geq 0$, with β a *negative* number, and let A

be its infinitesimal generator.   We shall prove the following:

Theorem 1.1: *Given an almost-periodic function* $f(t)$ *,* $t \in R \to X$ *, such that* $f(t) \in C^1(R;X)$ *and* $\sup_{t \in R} \| f'(t) \| < \infty$ *, there exists one and only one (strong) solution* $u(t)$ *of the equation* $u'(t) = Au(t) + f(t)$ *over* $t \in R$ *, which is almost-periodic.*

Proof of unicity: We shall use the elementary fact ([3], [6]) that almost-periodic functions are bounded over the real line.   Hence, it is sufficient to prove that $u(t) \equiv \theta$ is the only bounded solution over the real line of the homogeneous equation $u' = Au$, and this is a corollary of Theorem 1.1 in Chapter V.

Proof of existence: Let us consider the function $u(t)$ defined by the following (improper) integral:

$$u(t) = \int_{-\infty}^{t} S(t-\sigma)f(\sigma)d\sigma = \lim_{R \downarrow -\infty} \int_{R}^{t} S(t-\sigma)f(\sigma)d\sigma \ .$$

The function $S(t-\sigma)f(\sigma)$ is continuous if $R \leq \sigma \leq t$ with respect to σ
(if $\sigma_n \to \sigma_o \in [R,t]$, $S(t-\sigma_n)f(\sigma_n) - S(t-\sigma_o)f(\sigma_o) = S(t-\sigma_n)[f(\sigma_n)-f(\sigma_o)]$
$+ [S(t-\sigma_n) - S(t-\sigma_o)]f(\sigma_o)$ and $\| S(t-\sigma_n)f(\sigma_n) - S(t-\sigma_o)f(\sigma_o) \|$
$\leq Me^{\beta(t-\sigma_n)} \| f(\sigma_n)-f(\sigma_o) \| + \| S(t-\sigma_n)f(\sigma_o)-S(t-\sigma_o)f(\sigma_o) \| \leq M\| f(\sigma_n)-f(\sigma_o) \| + o(1)$,
$n \to \infty)$.

Hence the integral (in Riemann's sense) $\int_{R}^{t} S(t-\sigma)f(\sigma)d\sigma$ has a sense.
The integrand is estimated by:

$$\| S(t-\sigma)f(\sigma) \| \ \leq \ Me^{\beta(t-\sigma)} \cdot \sup_{\sigma \in R} \| f(\sigma) \| \ = \ Ce^{\beta t} \ e^{|\beta|\sigma} \ ;$$

also

$$\int_{-\infty}^{t} e^{|\beta|\sigma} d\sigma = \frac{1}{|\beta|} e^{|\beta|t} = \frac{1}{|\beta|} e^{-\beta t}$$

is convergent; thus, the integral

$$\int_{-\infty}^{t} S(t-\sigma)f(\sigma)d\sigma$$

is absolutely convergent and

$$\left\| \int_{-\infty}^{t} S(t-\sigma)f(\sigma)d\sigma \right\| \leq \frac{M}{|\beta|} \cdot \sup_{\sigma \in R} \| f(\sigma) \| \quad .$$

Actually, a very similar argument proves almost-periodicity of $u(t)$; given $\varepsilon > 0$, we consider, in view of the almost-periodicity of f , a number $L(\frac{\varepsilon|\beta|}{M})$ such that in any interval $[a, a+L]$ one finds a number $\tau(\frac{\varepsilon|\beta|}{M})$ with property that: $\displaystyle\sup_{t \in R} \| f(t+\tau) - f(t) \| < \frac{\varepsilon|\beta|}{M}$.

Afterwards, we can write:

$$u(t+\tau)-u(t) = \int_{-\infty}^{t+\tau} S(t+\tau-\sigma)f(\sigma)d\sigma - \int_{-\infty}^{t} S(t-\sigma)f(\sigma)d\sigma$$

$$= \int_{-\infty}^{t} S(t-\xi)\,[f(\xi+\tau)-f(\xi)]d\xi \qquad (\text{when} \quad \sigma-\tau = \xi)$$

and consequently: $\displaystyle \| u(t+\tau)-u(t) \| \leq \frac{\varepsilon|\beta|}{M} \int_{-\infty}^{t} Me^{\beta(t-\xi)}d\xi = \varepsilon ,$ $\quad t \in R$, thus

proving the almost-periodicity of $u(t)$.

(The strong continuity of $u(t)$ can be proved in a similar way using uniform continuity on the real line of any almost-periodic function ([3]), [6]): given $\varepsilon > 0$, $\exists\, \delta(\varepsilon)$ such that $|h| < \delta(\varepsilon) \Rightarrow \displaystyle\sup_{t \in R} \| f(t+h)-f(t) \| < \varepsilon;$ thus, $\| u(t+h)-u(t) \| \leq \varepsilon \frac{M}{|\beta|}$ for $|h| < \delta(\varepsilon)$, $t \in R$).

It remains to show that the above defined function $u(t)$ is a strong solution on R of : $u' - Au = f$. We have: $u(t) = \lim_{n\to\infty} u_n(t)$, where $u_n(t) = \int_{-n}^{t} S(t-\sigma)f(\sigma)d\sigma$ for $t > -n$. We shall first establish differentiability of $u_n(t)$ in the following way: It is:

$$u_n(t) = \int_{-n}^{t} S(t-\sigma)f(\sigma)d\sigma = \int_{0}^{t+n} S(u)f(t-u)du$$

$$= \int_{0}^{x} S(u)f(x-n-u)du = v_n(x) = \int_{0}^{x} S(u)g_n(x-u)du$$

where $g_n(x) = f(x-n)$ is a function in $C^1(R;X)$. According to Theorem 5.1 in Chapter II we find that (after the remark that $x = t + n > 0$), $v_n'(x)$ exists, strongly, for $x > 0$, $v_n(x) \in D(A)$, $x > 0$, $v_n'(x) = Av_n(x) + g_n(x)$, $x > 0$. Remark that $v_n(x) = v_n(t+n) = u_n(t)$; we see that the equality:

$u_n'(t) = Au_n(t) + g_n(t+n) = Au_n(t) + f(t)$, is verified.

On the other hand, if we remember the proof of Theorem 5.1 in Chapter II, we see that $v_n'(x) = \int_{0}^{x} S(u)g_n'(x-u)du + S(x)g_n(0)$, and therefore

$$u_n'(t) = \int_{0}^{t+n} S(u)f'(t-u)du + S(t+n)f(-n) .$$

Using estimate: $\| S(t+n)f(-n) \| \leq Me^{\beta t} e^{\beta n} \cdot \sup_{t \in R} \| f(t) \|$ and also:

$\sup_{\xi \in R} \| f'(\xi) \| < \infty$ we derive that $\lim_{n\to\infty} u_n'(t)$ exists for any $t \in R$ and

$$= \int_{0}^{\infty} S(u)f'(t-u)du .$$

Let us take now any compact interval $[a,b] \subset R$; then choose n , such that $-n < a$. For $t \in [a,b]$ we can write: $u_n(t) = u_n(a) + \int_{a}^{t} u_n'(s)ds$.

Remark also that $u_n'(t) \to v(t) = \int_0^\infty S(u)f'(t-u)du$ *uniformly on* $[a,b]$

(In fact: $\|u_n'(t) - v(t)\| \le \int_{t+n}^\infty \|S(u)f'(t-u)\| du + Me^{\beta t} e^{\beta n} \sup_{t\varepsilon R}\|f(t)\| \le$

$(\int_{t+n}^\infty Me^{\beta u} du)\sup_{\xi\varepsilon R}\|f'(\xi)\| + (Me^{\beta a} \sup_{t\varepsilon R}\|f(t)\|)e^{\beta n} \le (\frac{c}{|\beta|}e^{\beta a})e^{\beta n} + c_1 e^{\beta n} =$

$c_2 e^{\beta n} \to 0$ as $n \to \infty$ (therefore, uniformly for $t \varepsilon [a,\infty)$).

At this point we use the relation: $u_n(t) = u(a) + \int_a^t u_n'(s)ds$, $a < t$.

From the obvious continuity of any derivative $u_n'(t)$ we derive continuity

of $v(t)$; also, we obtain, as $n \to \infty$, $u(t) = u(a) + \int_a^t v(s)ds$, $t > a$.

Thus, $u(t)$ has a strong derivative $= v(t)$ which is strongly continuous.

Consider now the equality: $Au_n(t) = u_n'(t) - f(t)$. Remember that A is a

closed operator; if $n \to \infty$, $u_n(t) \to u(t)$, $Au_n(t) \to v(t) - f(t) = u'(t) - f(t)$

accordingly $u(t) \varepsilon D(A)$ and $Au(t) = u'(t) - f(t)$ for any $t \varepsilon R$.

This ends the proof of Theorem 1.1.

<u>Corollary</u>: Under the assumption of the Theorem, any bounded over R

solution of equation $u' = Au + f$ is almost-periodic.

In fact, we have proved that this equation has exactly one bounded solution,

hence it must coincide with $\int_{-\infty}^t S(t-\sigma)f(\sigma)d\sigma$ and is almost-periodic.

<u>Remark</u>: If $f(t)$ is not almost-periodic but belongs to $C^1(R;X)$, is

uniformly bounded over R and has an uniformly bounded derivative over R ,

the above proof shows that $u(t) = \int_{-\infty}^t S(t-\sigma)f(\sigma)d\sigma$ is the (unique) *bounded*

over R strong solution of the equation $u' = Au + f$.

References

1 N. I. Achieser and I. M. Glazman, Theorie der linearen Operatoren im
   Hilbert-Raum, Akademie-Verlag, Berlin 1954.

2 S. Agmon and L. Nirenberg, Properties of solutions of ordinary
   differential equations in Banach space,
   Comm. Pure Appl. Math., vol XVI, No 2 (1963),
   pp 121-239.

3 L. Amerio and G. Prouse, Almost-periodic functions and functional
   equations, Van Nostrand-Reinhold Co.,
   New York - Toronto - London - Melbourne,
   1971.

4 P. Arminjon, Existence et régularité de solutions
   élémentaires d'un opérateur differentiel
   abstrait, Math. Ann. 187 (1970), pp 117-126.

5 J. Barros-Neto, An introduction to the Theory of Distributions,
   Marcel Dekker, Inc., New York 1973.

6 S. Bochner, Abstrakte fast-periodische Funktionen, Acta
   Math., 61 (1933), pp 149-184.

7 R. Carroll, Abstract methods in partial differential
   equations, Harper and Row Publishers,
   New York - Evanston - London, 1969.

8 D. L. Colton, Partial differential equations in the
   complex domain, Pitman Publishing, London -
   San Francisco - Melbourne, 1976.

9 M. G. Crandall and A. Pazy, On the differentiability of weak solutions
   of a differential equation in a Banach
   space, J. Math. Mech. 10 (1969),
   pp 1007-1016.

10 N. Dunford and J. T. Schwartz, Linear operators, Part I, Interscience
   Publishers Inc., New York 1958.

11 A. Friedman, Partial Differential Equations, Holt,
   Rinehart and Winston Inc., New York -
   Montreal - Toronto - London, 1969.

12 I. Gelfand, On one-parametrical groups of operators in
   a normed space, C. R. (Doklady) Acad. Sci.
   U.R.S.S., NS 25 (1939), pp 713-718.

13 S. Goldberg, Unbounded Linear Operators (Theory and
   Applications), McGraw-Hill Book Company,
   New York - Toronto-London, 1966.

14   J. Goldstein,   On the growth of solutions of inhomogeneous
                                    abstract wave equations, Journal of Math.
                                    Analysis and Appl., vol. 37 (1972),
                                    pp 650-654.

15   E. Hille,   Functional analysis and Semi-groups,
                                    American Mathematical Society Colloquium
                                    Publications, vol XXXI, New York, 1948.

16   E. Hille and R. S. Phillips,   Functional analysis and Semi-Groups, AMS
                                    Coll. Publ., (revised edition), vol. XXXI,
                                    Providence (R.I.), 1957.

17   M. W. Hirsch and S. Smale,   Differential equations, Dynamical Systems
                                    and Linear Algebra, Academic Press,
                                    New York - San Francisco - London, 1974.

18   L. Hörmander,   Linear partial differential operators
                                    (Third revised printing) Springer-Verlag,
                                    New York Inc. 1969.

19   A. N. Kocubei,   Fundamental solutions of differential-
                                    operational equations (Russian),
                                    Differencialnie Uravnenia, Tom XIII, no 9,
                                    1977, pp 1588-1597.

20   S. G. Krein,   Linear Differential Equations in Banach
                                    Space (Translations of Mathematical
                                    Monographs) American Mathematical Society
                                    Providence, R.I., 1971.

21   G. Ladas and V. Lakshmikantham, Differential equations in abstract spaces,
                                    Academic Press, New York 1972.

22   P. D. Lax,   A stability theorem for solutions of abstract
                                    differential equations and its applications
                                    to the study of the local behavior of
                                    solutions of elliptic equations, Comm. Pure
                                    Appl. Math, 9 (1956) pp 747-766.

23   H. Levine,   On the uniqueness of bounded solutions to
                                    $u'(t) = A(t)u(t)$ and $u''(t) = A(t)u(t)$ in
                                    Hilbert space, SIAM J. Math. Anal. 4 (1973),
                                    pp 250-259.

24   V. E. Liance,   A boundary value problem for parabolic
                                    systems of differential equations with
                                    strongly elliptic right-hand side (Russian),
                                    Matem. Sbornik, vol 35(77), 1954, pp 357-368.

25 B. Malgrange,   Existence et approximation des solutions
                                    des équations aux dérivées partielles et
                                    des équations de convolution, Ann. Inst.
                                    Fourier Grenoble, 6 (1955-56), pp 3-86.

26 B. Malgrange,   Operatori differenziali, C.I.M.E. (Teoria
                                    delle distribuzioni),, Edizioni Cremonese,
                                    Roma, 1961.

27 M. A. Malik,   Regularity of elementary solutions of
                                    abstract differential equations, Boll. U.M.I.
                                    (4) 4 (1971), pp 78-84.

28 M. A. Malik,   An approximation property for abstract
                                    differential equations, Rend. Sem. Mat. Univ.
                                    Padova, vol. 55 (1976), pp 45-48.

29 R. H. Martin, Jr.,   Nonlinear operators and differential
                                    equations in Banach spaces, John Wiley and
                                    Sons, New-York - London - Sydney - Toronto,
                                    1976.

30 E. Nelson,   Analytic vectors, Ann. of Math., 70 (1959),
                                    pp 572-615.

31 N. Pavel,   Sur certaines équations différentielles
                                    abstraites, Boll. Un. Mat. Ital. (4)
                                    6 (1972), pp 397-409.

32 A. S. Rao and W. Hengartner,   On the existence of a unique almost periodic
                                    solution of an abstract differential
                                    equation, J. London Math. Soc. (2) 8 (1974),
                                    pp 577-581.

33 M. Reed and B. Simon,   Methods of modern mathematical physics, I,
                                    II, Academic Press, New York and London,
                                    1972 and 1975.

34 F. Riesz and B. von Sz Nagy,   Lecons d'Analyse Fonctionnelle, Akad. Kiado,
                                    Budapest, 1952.

35 S. Rosencrans,   Growth of solutions of a differential
                                    equation in Hilbert space, The Proceedings
                                    of the Louisiana Academy of Sciences, vol.
                                    XXXI, Dec. 1969, p. 98.

36 S. Sobolev,   Applications of Functional Analysis in
                                    Mathematical Physics, American Mathematical
                                    Society, Translations of Mathematical
                                    Monographs, vol. 7, Providence, R.I., 1963.

37 K. Yosida,   Functional Analysis, Springer-Verlag,
                                    Berlin-Güttingen - Heidelberg, 1965.

38   S. Zaidman,   Sur une application du Théoreme de Hille-Yosida, IV Congres des mathématiciens roumains, 1956, Editura Academiei R.P.R., 1960.

39   S. Zaidman,   Sur la perturbation presque-périodique des groupes et semi-groupes de transformations d'un espace de Banach, Rend. Matem. e sue Appl., S.V., 16 (1957) pp 197-206.

40   S. Zaidman,   Sur la presque-périodicité des solutions de l'équation des ondes non homogène, Journ. Math. Mech., vol. 8 (1959) pp 369-382.

41   S. Zaidman,   A global existence theorem for some differential equations in Hilbert spaces, Proc. Nat. Acad. Sci. USA, vol 51, no 6, 1964, pp 1019-1022.

42   S. Zaidman,   Equations différentielles abstraites, Les Presses de l'Université de Montréal, Janvier 1966.

43   S. Zaidman,   On elementary solutions of abstract differential equations, Boll. U.M.I. (4), N. 4-5 (1969), pp 487-490.

44   S. Zaidman,   Bounded solutions of some abstract differential equations, Proc. Amer. Math. Soc., vol. 23, no 2 (1969), pp 340-342.

45   S. Zaidman,   Uniqueness of bounded solutions for some abstract differential equations, Annali Univ. Ferrara, vol. XIV, 1969, pp 101-104.

46   S. Zaidman,   Some asymptotic theorems for abstract differential equations, Proc. Amer. Math. Soc., vol 25, no 3 (1970), pp 521-525.

47   S. Zaidman,   On a certain approximation property for first-order abstract differential equations, Rend. Sem. Mat. Univ. Padova, 46 (1971), pp 191-198.

48   S. Zaidman,   A generator of a strongly continuous semi-group, Notices A.M.S., 1971, Abstract 71 T-B161, pag. 809.

49   S. Zaidman,   Problem 5961*, American Math. Monthly, 1974, p. 293.